SEVEN MILES DEEP

MINING FACES FROM THE OWEN FITZGERALD COLLECTION

THE BEATON INSTITUTE

Cape Breton University Press
Sydney, Nova Scotia

Cape Breton University Press recognizes the support of the Province of Nova Scotia, through the Department of Tourism, Culture and Heritage. We are pleased to work in partnership with the Culture Division to develop and promote our cultural resources for Nova Scotians. We acknowledge the support of the Canada Council for the Arts for our publishing program.

NOVA SCOTIA
Tourism, Culture and Heritage

Canada Council for the Arts Conseil des Arts du Canada

Cover design Cathy MacLean, Pleasant Bay, NS.
Layout by Mike Hunter, Louisbourg, NS.
Printing by Kromar Printing Ltd. Winnipeg, Manitoba
Printed in Canada.

Library and Archives Canada Cataloguing in Publication

Beaton Institute of Cape Breton Studies
Seven miles deep : mining faces from the Owen Fitzgerald collection / Owen Fitzgerald, the Beaton Institute.

ISBN 978-1-897009-25-3

1. Fitzgerald, Owen. 2. Coal miners--Nova Scotia--Cape Breton Island--Portraits. 3. Coal mines and mining--Nova Scotia--Cape Breton Island--Pictorial works. 4. Portrait photography--Nova Scotia--Cape Breton Island. I. Fitzgerald, Owen II. Title.

TR681.M64B42 2007 779'.9622334097169092 C2007-906244-X

Cape Breton University Press
PO Box 5300
Sydney, NS B1P 6L2 Canada

SEVEN MILES DEEP

MINING FACES FROM THE OWEN FITZGERALD COLLECTION

FOREWORD

When asked, they say they are miners; sharing the common bond of occupation, but below the surface their individuality emerges. They were and are electricians, millwrights, machinists and mechanics working in the dark of the mines. Images included here capture a range of mines including Phalen Mine, Lingan Mine, Prince Mine, Donkin Mine and No. 26 Colliery between 1974 and 1993. There is one photograph from Princess Mine (1975).

Many above ground know of the importance of the mining industry, but few have the opportunity to glimpse that world. Through the camera lens of Owen Fitzgerald this book illuminates the faces of the men who worked the mines. Many of these underground mining photos are unique and now historic; they are of miners, not just the geology, not just machinery. The pictures capture the context, conditions and camaraderie of the work—the environment could be cold with the rush of underground winds, but the smiles exude the warmth of friendship.

These pictures come together as a collection as they focus on the human face of mining—which so deeply has and continues to influence the social and cultural community of Cape Breton Island. The faces printed here remain nameless for fear of leaving out or misrepresenting even one. The pictures are supported by words when it may help to have context and texture—the text has been supported through the oral history collections of the Beaton Institute. Transcripts of taped interviews were examined to enrich the visual presentation. Owen Fitzgerald's recollection of his experiences with the miners and the taking of these photographs adds to this oral history. As well, a small group of miners were invited to preview the images and their reactions, information and enthusiasm are captured here. Their conversation has been recorded and added to the holdings of the Beaton Institute archives for future researchers.

It was not the purpose of this book to review the history of coal mining—that is left for others. Nor was it our task to critique the culture of mining, though the faces and voices of coal mining presented here contribute greatly to our knowledge. Past publications seemed to have focused on the machine, rather than the men. There are machines in these pictures, but they provide the backdrop for the personalities, camaraderie, strength and character of the men.

The goal of the book is to simply open and expose—if for but a moment—the Beaton Institute archive for all to see. It's been said often, and still holds true: a picture is worth a thousand words.

Catherine Arseneau
Manager, Beaton Institute

SEVEN MILES DEEP

In 1974, after spending three years in Toronto studying photography at Ryerson, my long-time attitude that Cape Breton has a lot going for it took on a whole new perspective.

With my Leica camera in hand, a renewed sense of pride and determination, I returned to my roots during what turned out to be an important chapter in Cape Breton's history. For the last quarter of the 20th century, I photographed Cape Breton and its people.

One of the most interesting and challenging photographic assignments I had was underground in the island's coal mines. It's another world, presenting many dramatic and unique photographic opportunities. The Cape Breton mines were submarine mining operations, as the mines extended up to seven miles out under the Atlantic Ocean.

I am very proud of my collection of these photographs. They have been published in *Time* magazine, *Maclean's*, *Reader's Digest* and in many books, including on the cover of several books on mining. I am delighted to see the publication of this collection of the faces of Cape Breton mining during an important chapter in our history. This collection results from dozens of trips out under the ocean to photograph what is a unique environment that few have seen or experienced.

When I went underground, I found a huge and alien community without light, yet with hundreds of men walking the miles of slopes and tunnels, working very hard, but in darkness: laughing, joking and with a comfort and pride that I didn't often see on the surface.

Taking photographs underground, with all that coal dust, was a real challenge. When I came out of the pit, it took several showers to get rid of all signs of that coal dust. Think what years of that coal dust must do to the miner's lungs.

From the mid 1970s to the mid 1990s, when I took most of my mining photos, there were four operating mines in Cape Breton, employing almost 5,000 people. The 70s were good times, as "Coal Was King" again, as a result of the 1973 oil crisis. During its hay day, during and prior to the Second World War, almost 20,000 men worked in the mines. In 2007, there are no working mines in Cape Breton, although Xstrata is investigating the possible reopening of the Donkin mine.

These miners are a unique breed of people. Their character, ruggedness and friendliness, shine through in the photographs. They take great pride in their work and think of themselves as a special group. They often say that if you haven't worked underground, you can't imagine what it's like and that's true. It is an underground community, almost with its own culture and rules, with hundreds of men working where the sun never shines.

Owen Fitzgerald, 2007

Beaton Institute, Owen Fitzgerald Collection
195-A-37 Princess Mine 1975

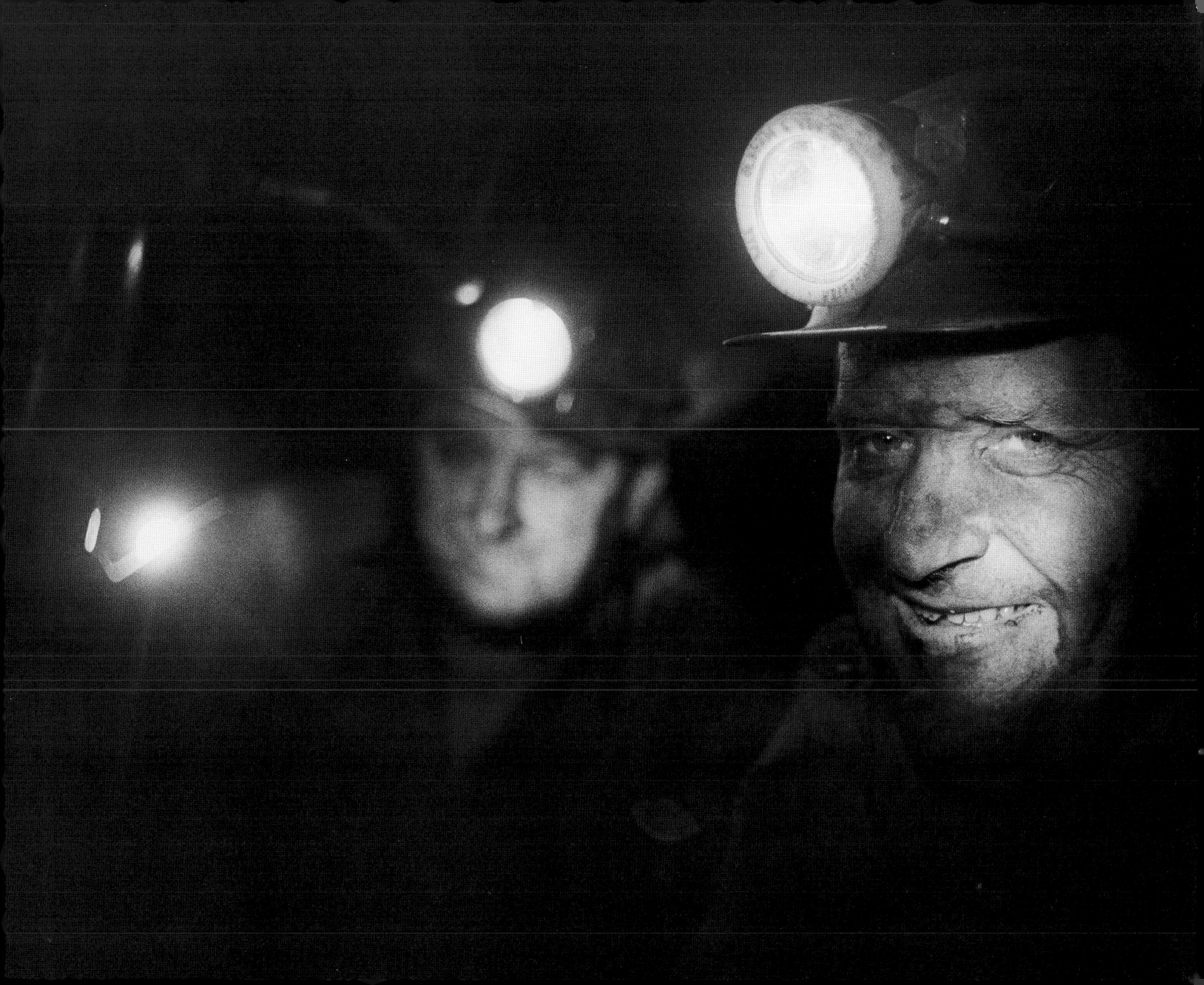

379-7-10A Phalen Mine, 1987

PHALEN MINE
1987
1988
1993

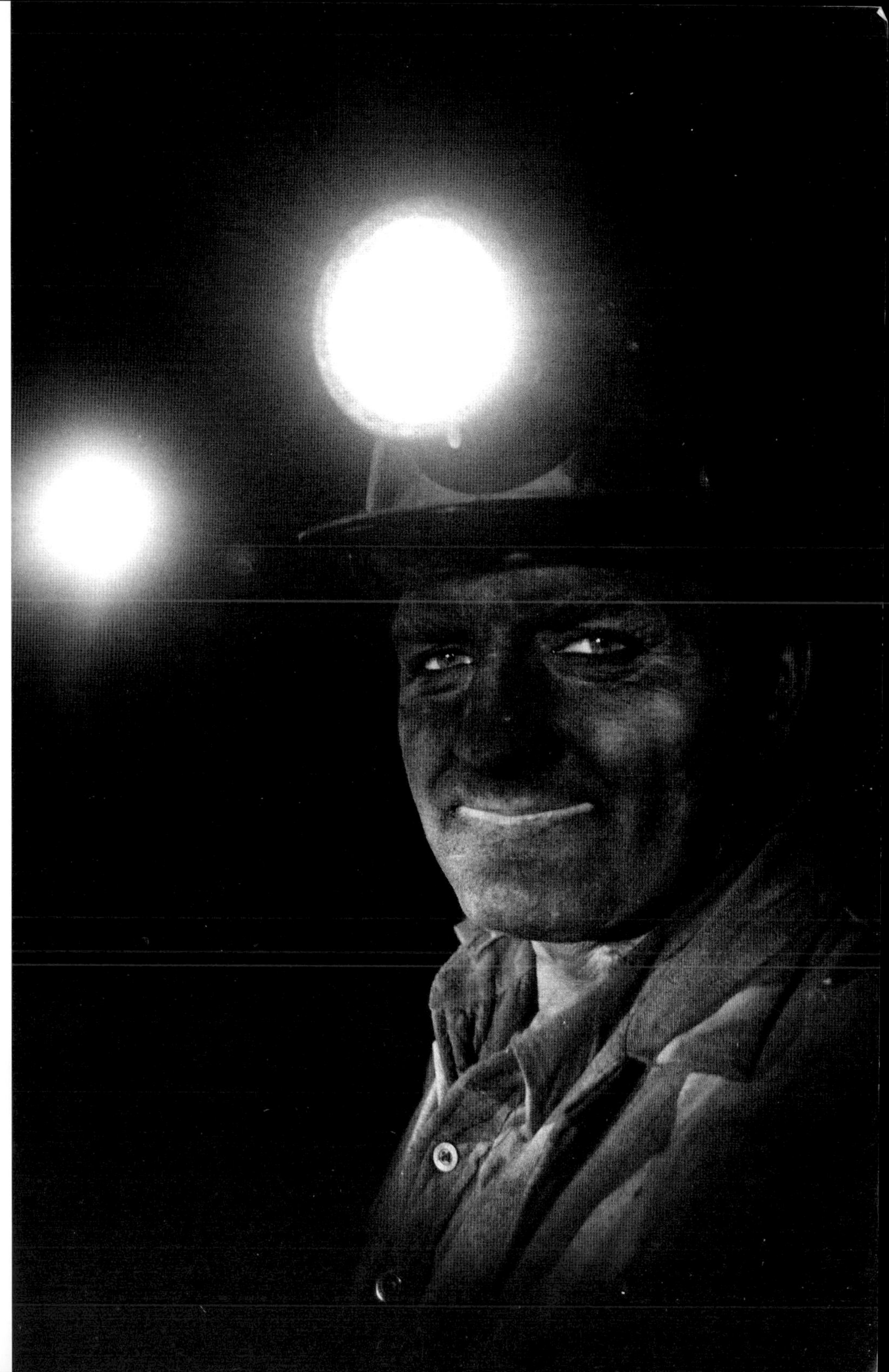

You went down in the man rake in the morning, came up in the evening.There was 8 or 9, we used to call them rakes on the chain, eh, 12 men to each rake. It took about 12-15 minutes to get down.

Beaton Institute Tape-2134 23 July 1984.

“Rake box” is an old English term, it was called a “trip” in the Prince Mine. This one had brakes and a convertable roof for access for a stretcher.

Right 85B-A-23 Phalen Mine 1993

Left 1000-A-06 Phalen Mine 1993

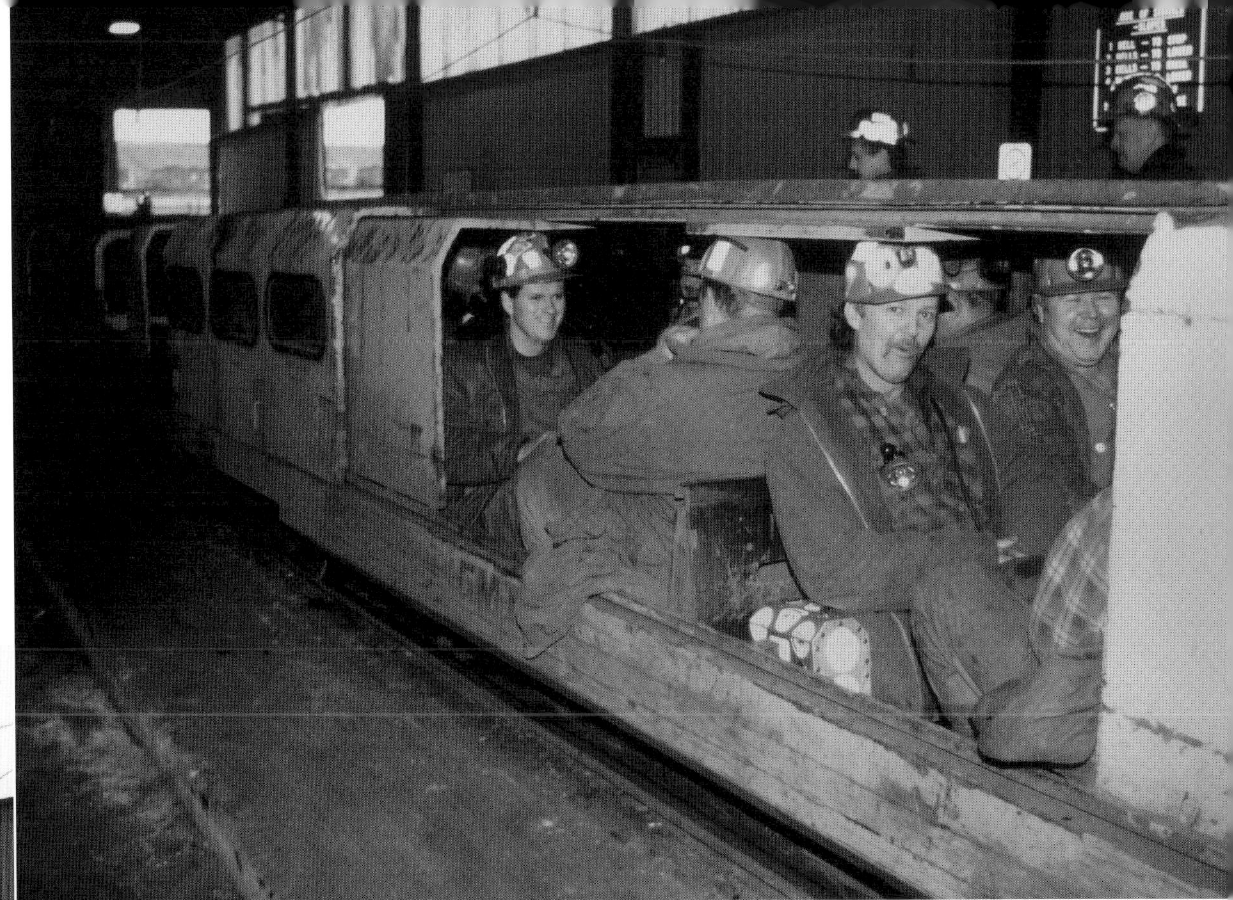

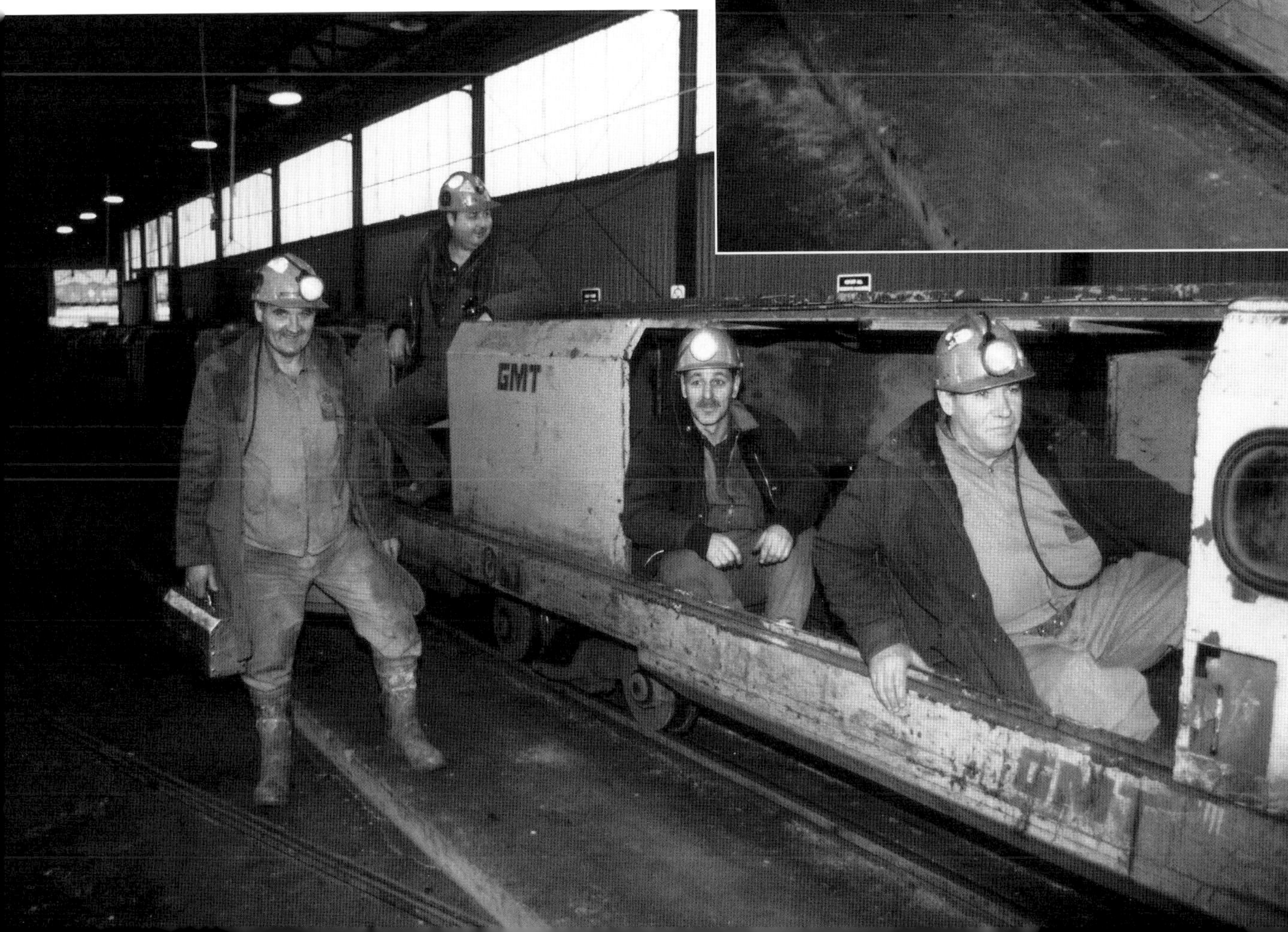
GMT

On one of my first trips underground, I was taken to the mine face or long-wall, where a large cutting machine, or shearer, would travel about a thousand feet up and down the length of a coal seam which ranged from 4 feet to 9 feet thick. At both ends of this cutting wall were tunnels going out to the main slope that went to the surface. Huge belts would carry the coal from the face to the surface.

The whole length of the cutting wall would be supported by huge hydraulic roof supports that would advance as the shearer cut deeper and deeper.

The miner that was showing me around the face failed to tell me about the "gob", and all of a sudden, there was a thunderous roar and I was sure that I was going to die, five miles out under the Atlantic Ocean. All the miners started to laugh and I felt like throwing my camera at them. I was then told that as the roof supports advance, the ceiling is left unsupported and as the pressure increases, the roof would, from time to time, come crashing down. This is referred to as the "Gob." No big deal, right!

Owen

A 500 hp Anderson shearer. Above are 2-legged jack roof supports used after 1987.

365-A-32 Phalen Mine, 1988

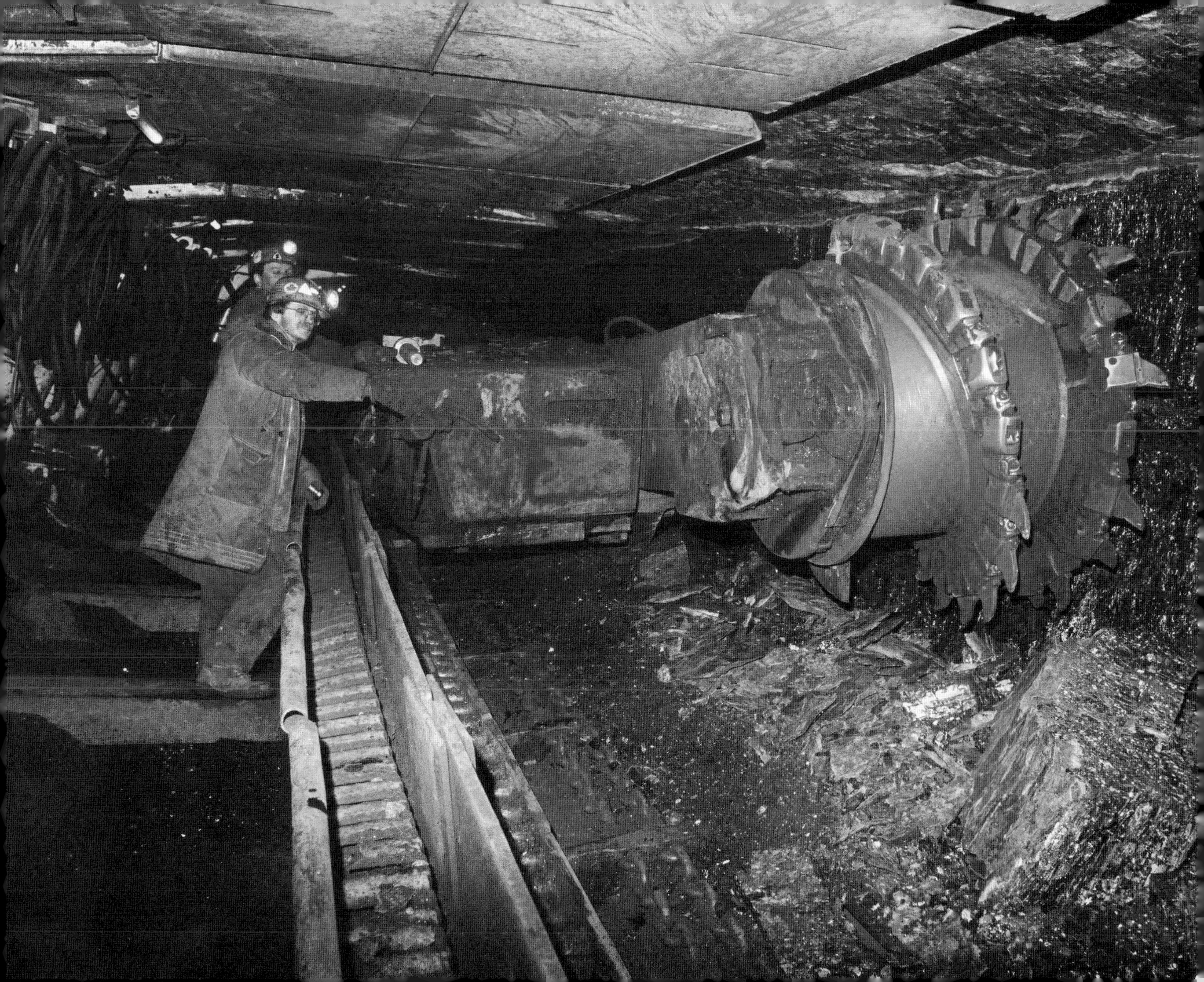

Constant wear on the shearer required frequent replacement of the picks (teeth) by the shear mechanic.

Inset: We went through thousands of vice-grips each year. We used to call them “New Waterford clothespins.”

379-6-3A Phalen Mine, 1987

Inset 365-B-10 Phalen Mine, 1988

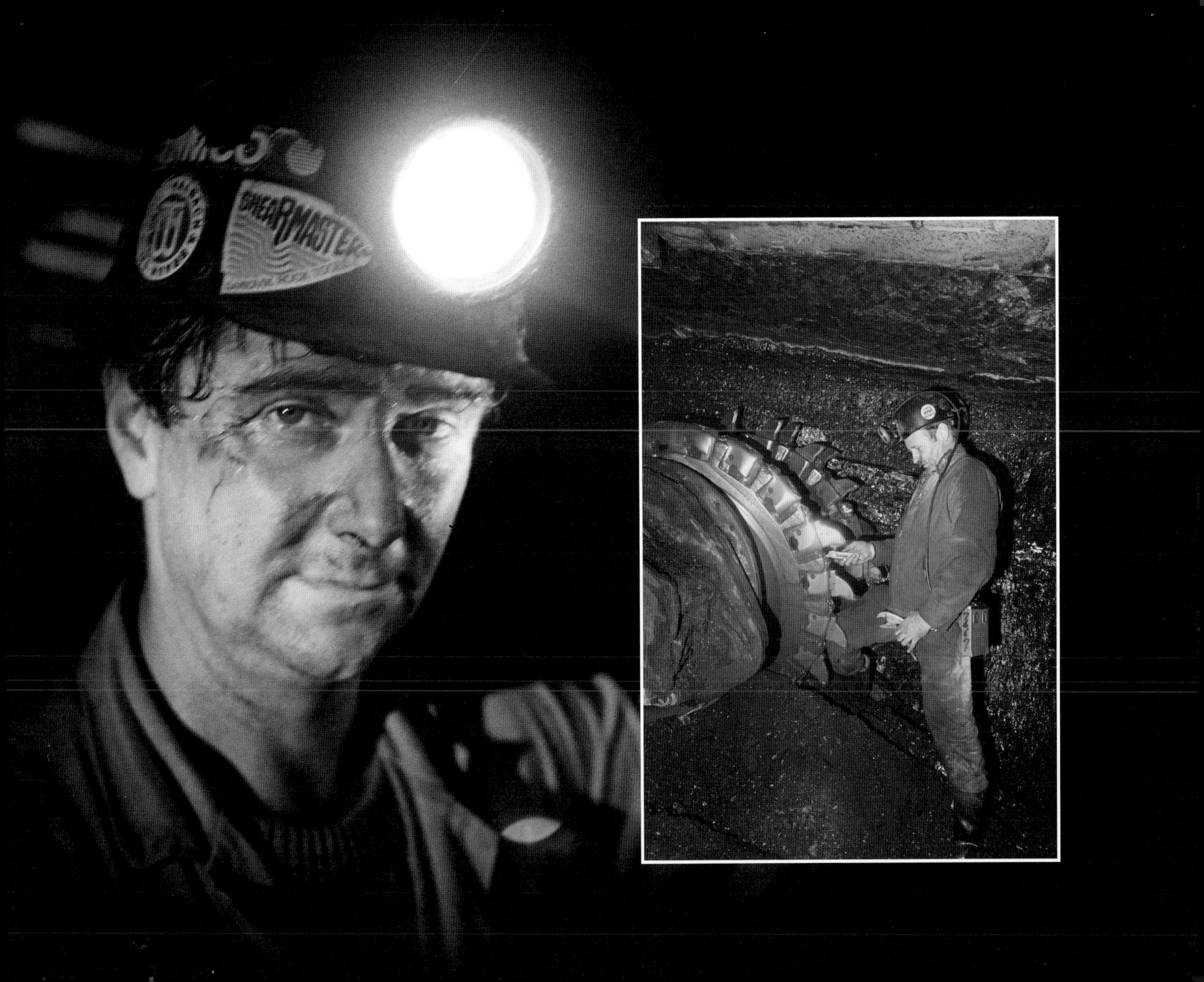

"I worked in the mines about 5 years and I got hurt. I was a machine runner that cuts the coal, and I was 7 months in the hospital … the machine kicked me in the guts, over the bladder. I'm still sore, that was in 1908. J. J. Chaisson.

Beaton Institute Tape-2141 27 July 1984.

Electricians and mechanics.

365-B-23 Phalen Mine, 1988

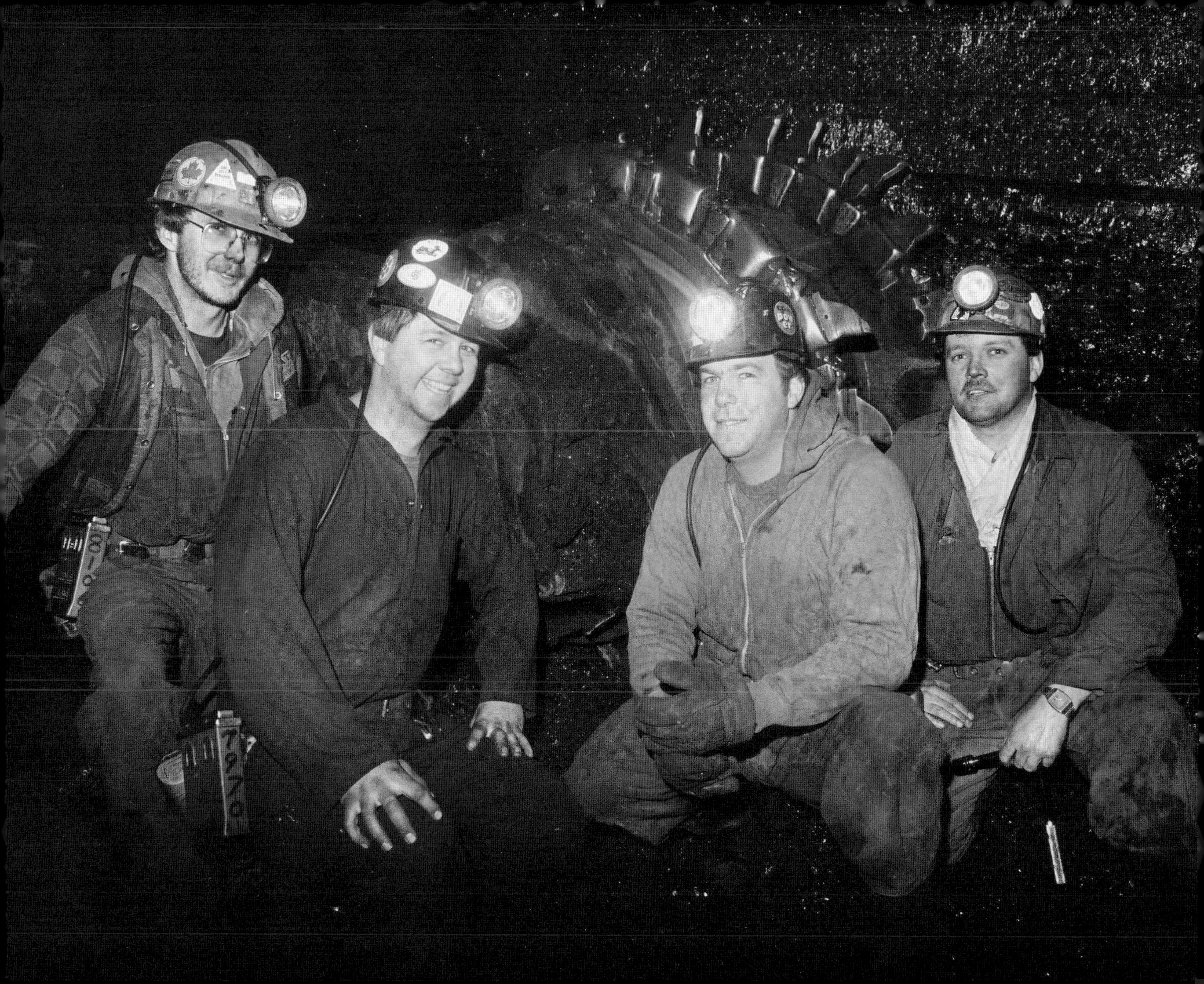

Immediately behind the shearer, large movable hydraulic roof supports hold up the ceiling. When the shearer is in operation, the noise is deafening and the dust choking. Getting around this working area is back-breaking, especially when lugging camera equipment along. Cape Breton miners would do this day after day. You are hunched over, sometimes on your knees, trying to get from one end of the long-wall to the other. The coal face, with its five-foot clearance, is a cramped and cluttered area, jammed with equipment. I can't imagine how they can get these large pieces of equipment into the mine. Apparently they are taken into the mine in pieces and assembled underground.

Owen

The mines continued to attract immigrants well into the 20th century.

Maintaining hydraulic roof supports.

On his belt is a "self-rescuer," a breathing apparatus which exchanges CO_2 for O_2.

365-A-8A Phalen Mine, 1988

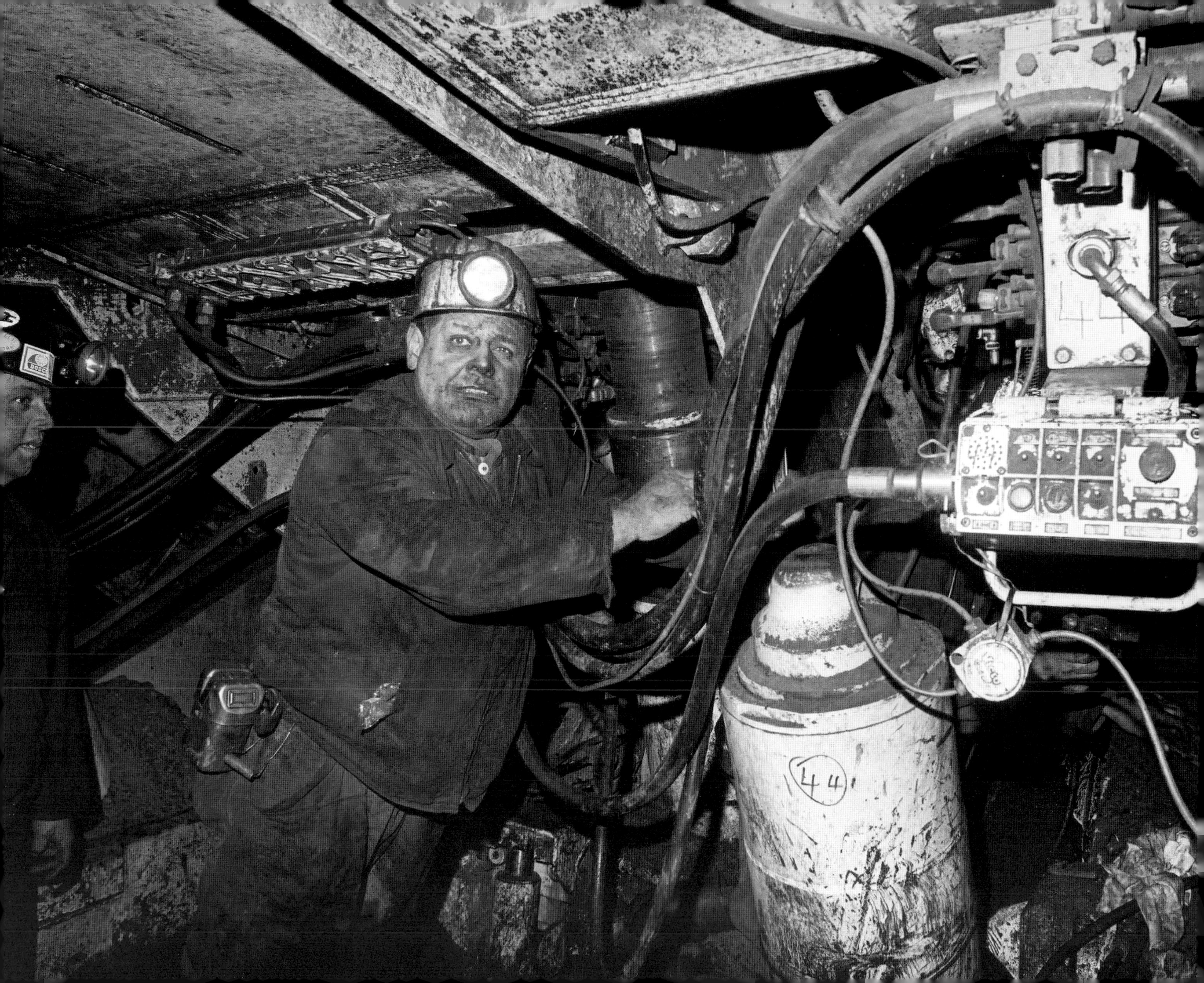
44

You can see on his lamp, he has his check number on his lamp ... that's his lamp. They were numbered sequentially.

365-A-7A Phalen Mine 1988

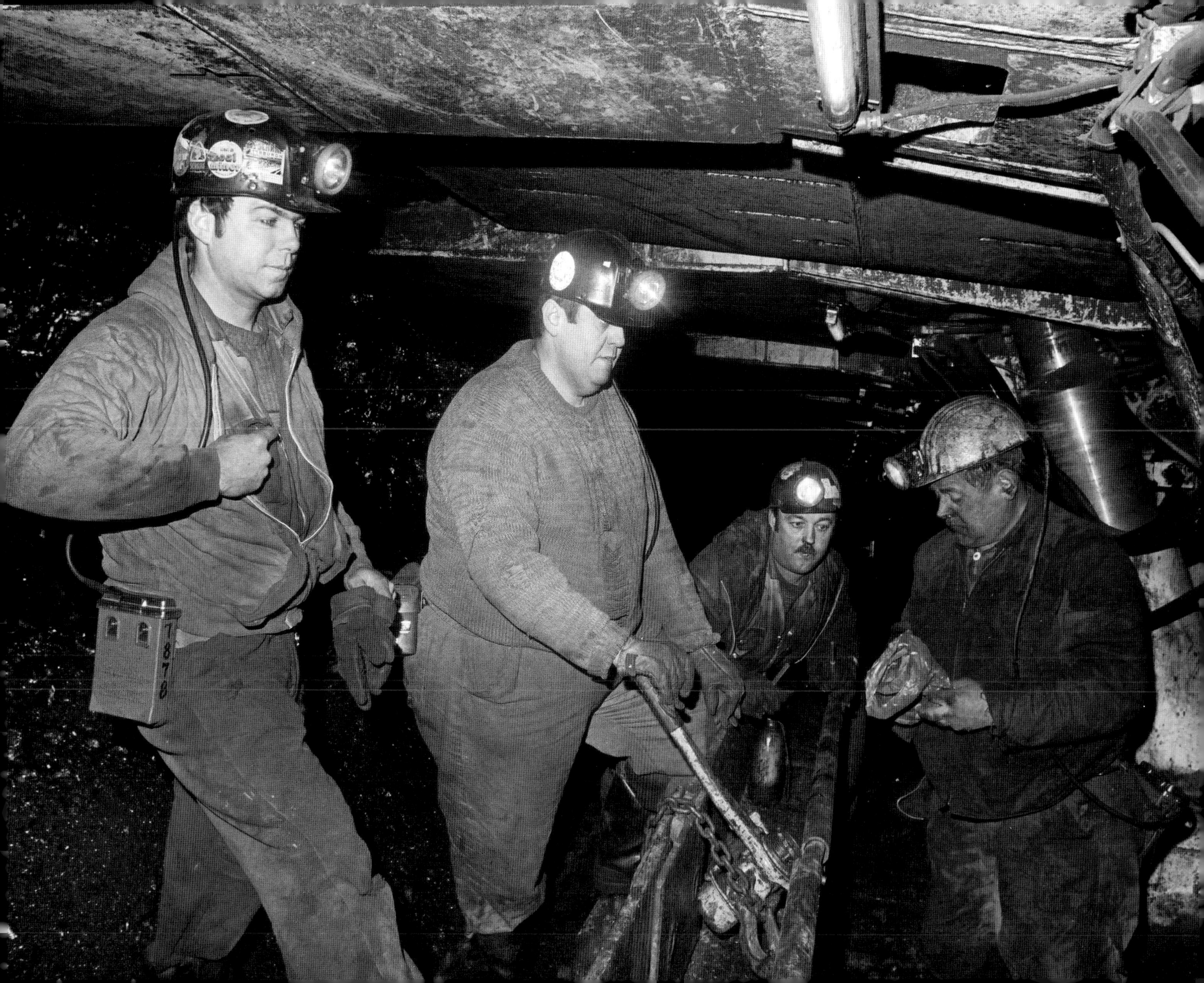
7878

Making my way up the long-wall, I choose every shot carefully, because the batteries in the explosion-proof flash only allow me 30-40 shots. I wrap my Nikon in a plastic bag and only expose it when I want to take a photograph, but still the dust is hard on the camera. More than once, I've had my camera seize up after spending a day underground. It was a costly, dirty, time consuming and backbreaking effort to get these photographs of Cape Breton coal miners working underground.

Many of my photographs are taken during temporary shutdowns of the shearer, because of the dust and the deafening noise and the fact that the miners are too busy working to be disturbed by a photographer and his blinding flash. Besides, it is just too dangerous when the shearer is operating, with stones flying, dust everywhere, noise, moving conveyer belts and the like.

Owen

365-B-35 Phalen Mine 1988

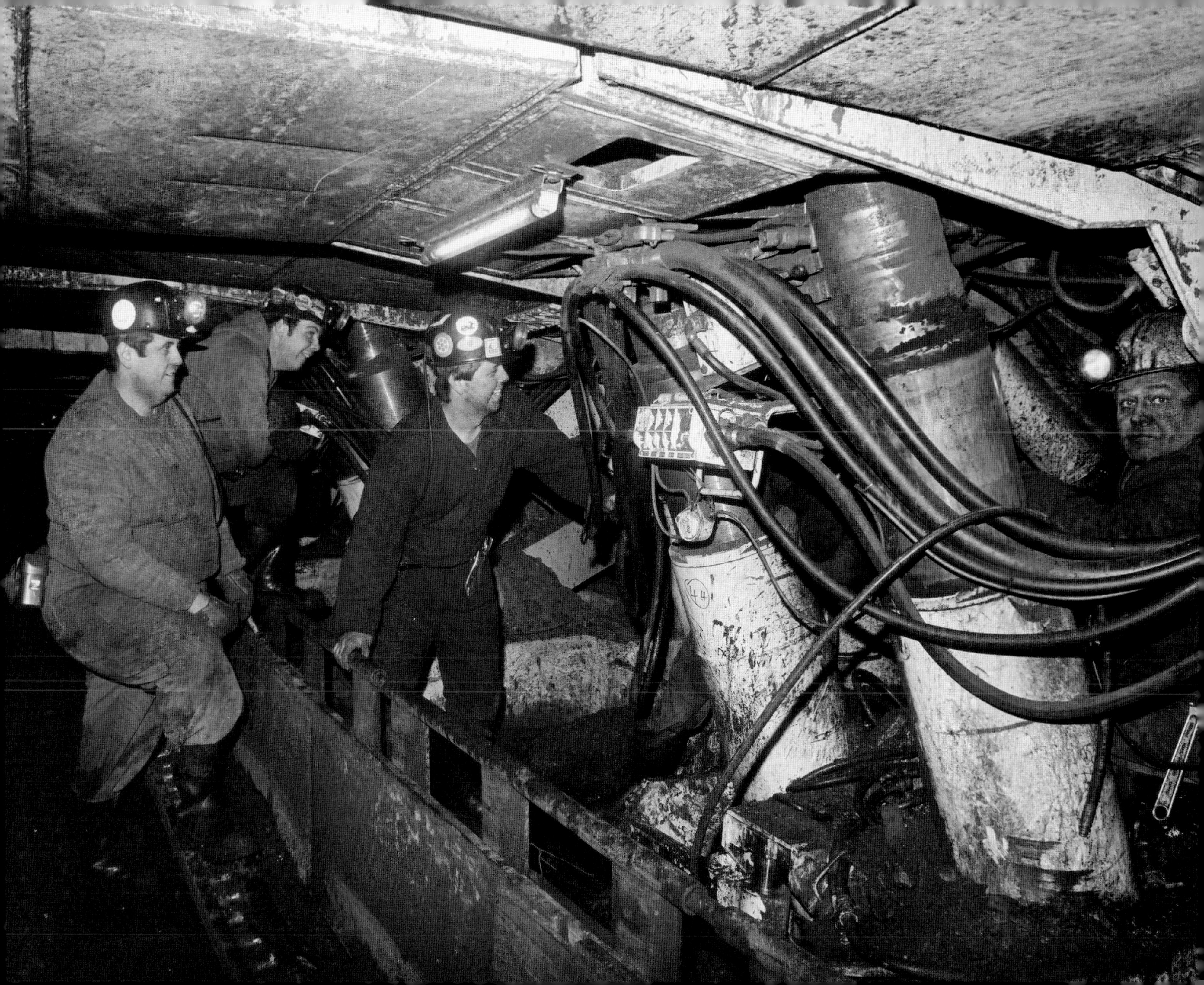

That's a roadheader. They use them today to drive subway tunnels. Her name was Angelina it was wrote right on the side of it.

An indication of the close relationship between man and machine in this environment, miners regularly referred to machinery by name, whether by function, manufacturer or a pet name.

Inset: He's got his coveralls rolled down. It was too warm down there sometimes.

379-4-15A Phalen Mine 1987

Inset 379-4-8A Phalen Mine 1987

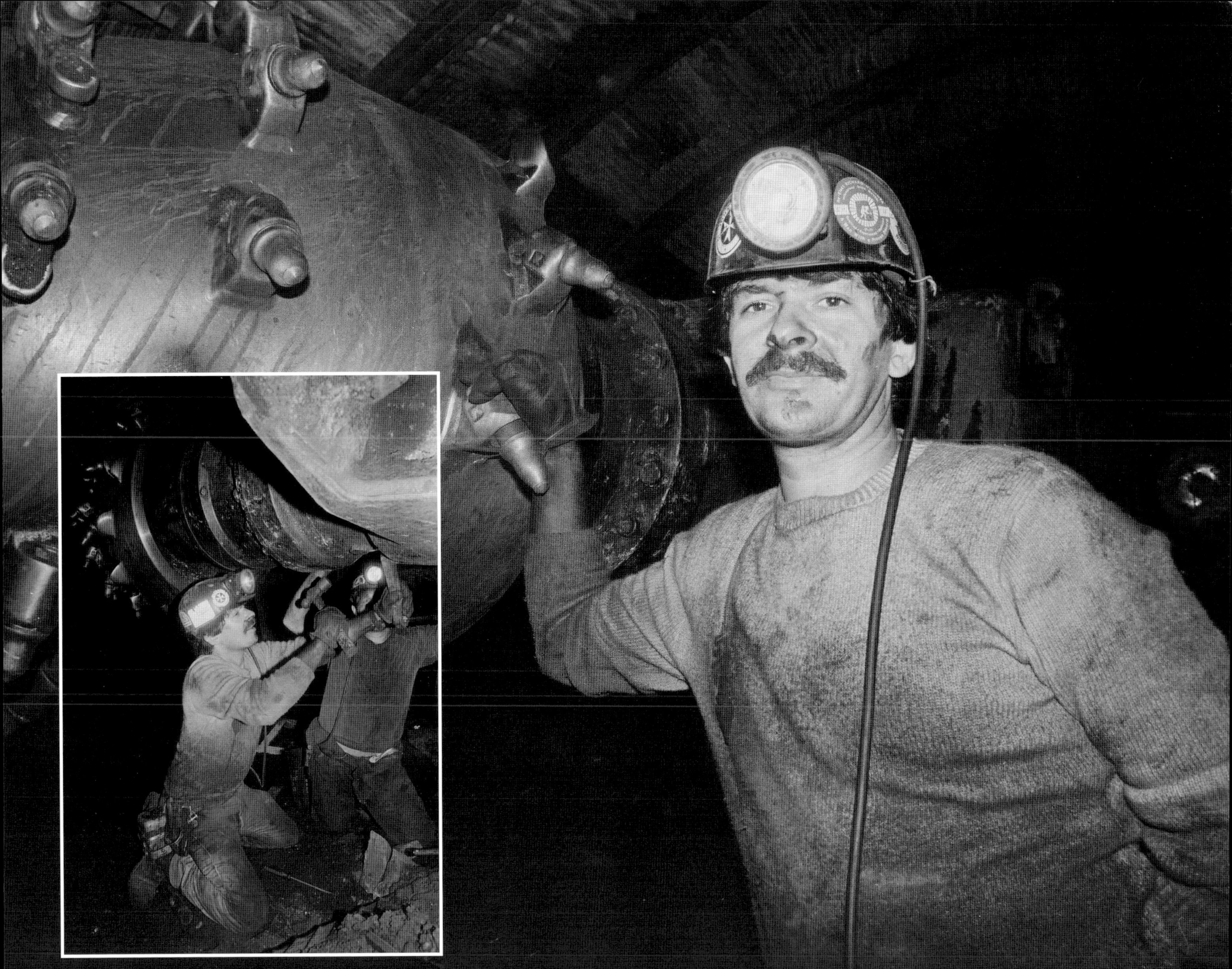

“I like mining, it wasn’t hard for me, I was pretty strong.
J. J. Chaisson

Tape-2141 27 July 1984.

379-4-11A Phalen Mine, 1987

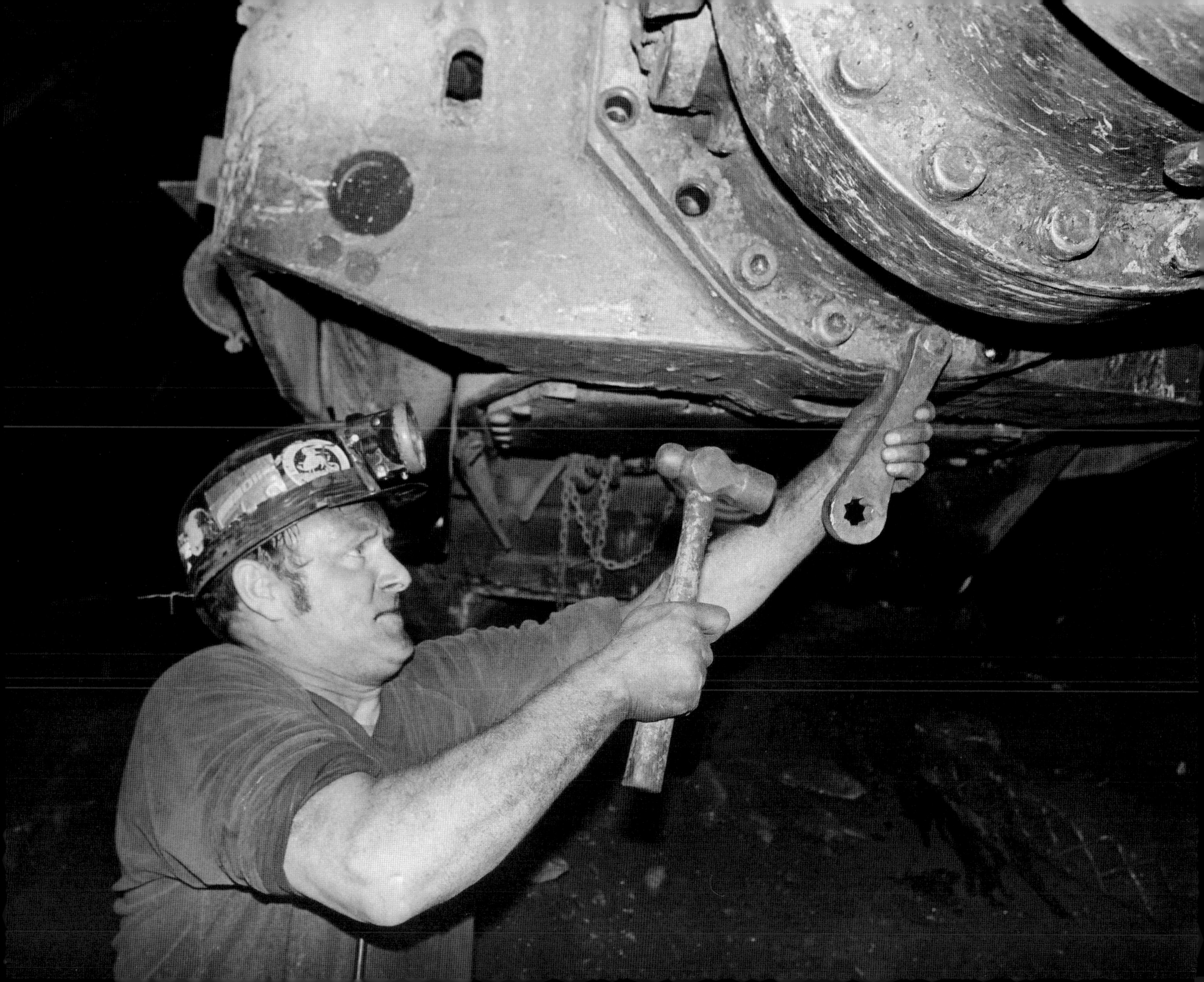

The underground manager was a great friend to me…He put me in a place … a good place … and I went in that good place, I made good money there.

Tape-2141 27 July 1984.

In the mine some parts were cold. One mine manager would ask if you needed a heater and then he handed you a shovel. Some of us called a shovel the Baldwin heater.

379-2-11 Phalen Mine 1987

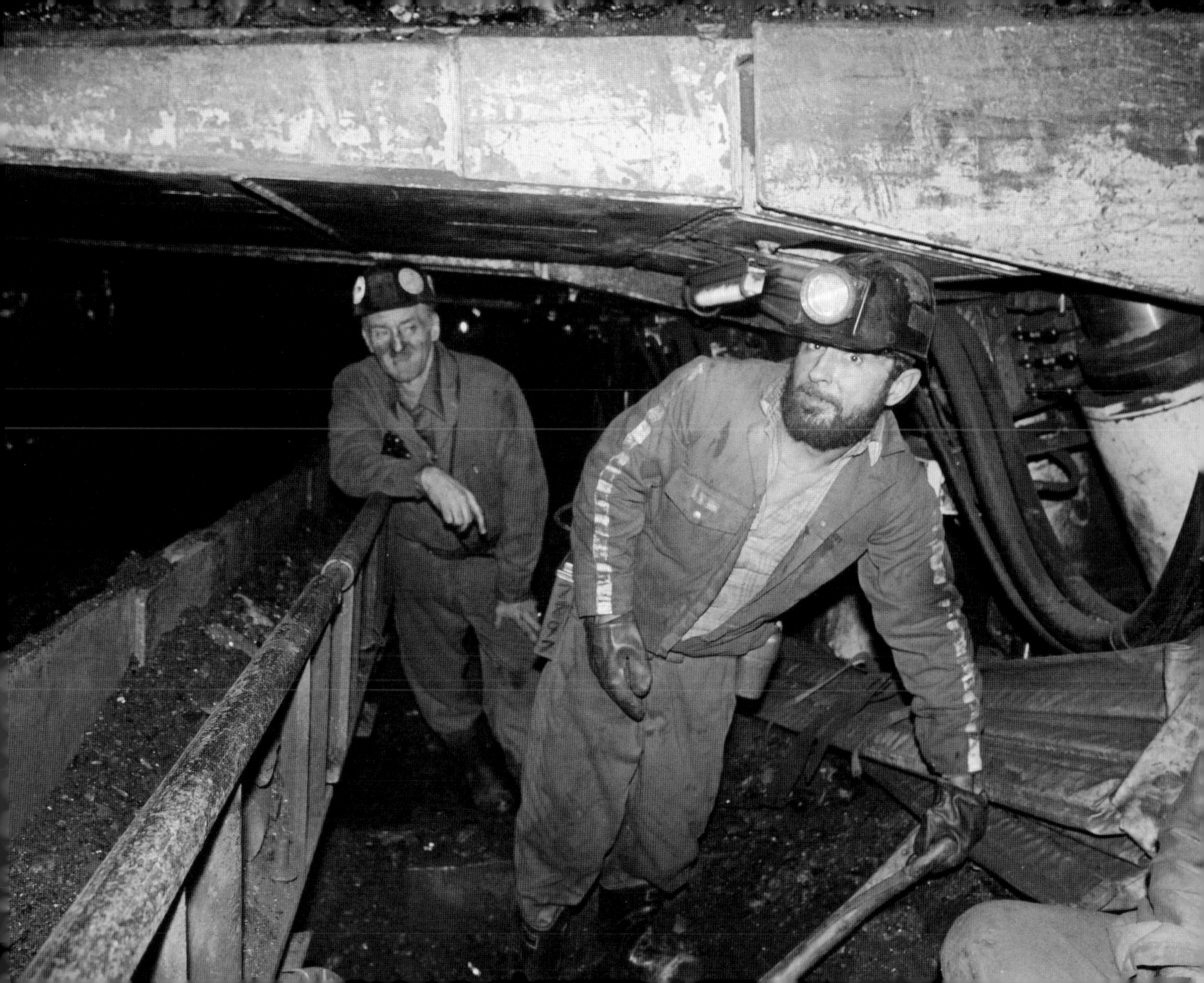

He’s probably wearing his old hat.

379-7-9A Phalen Mine 1987

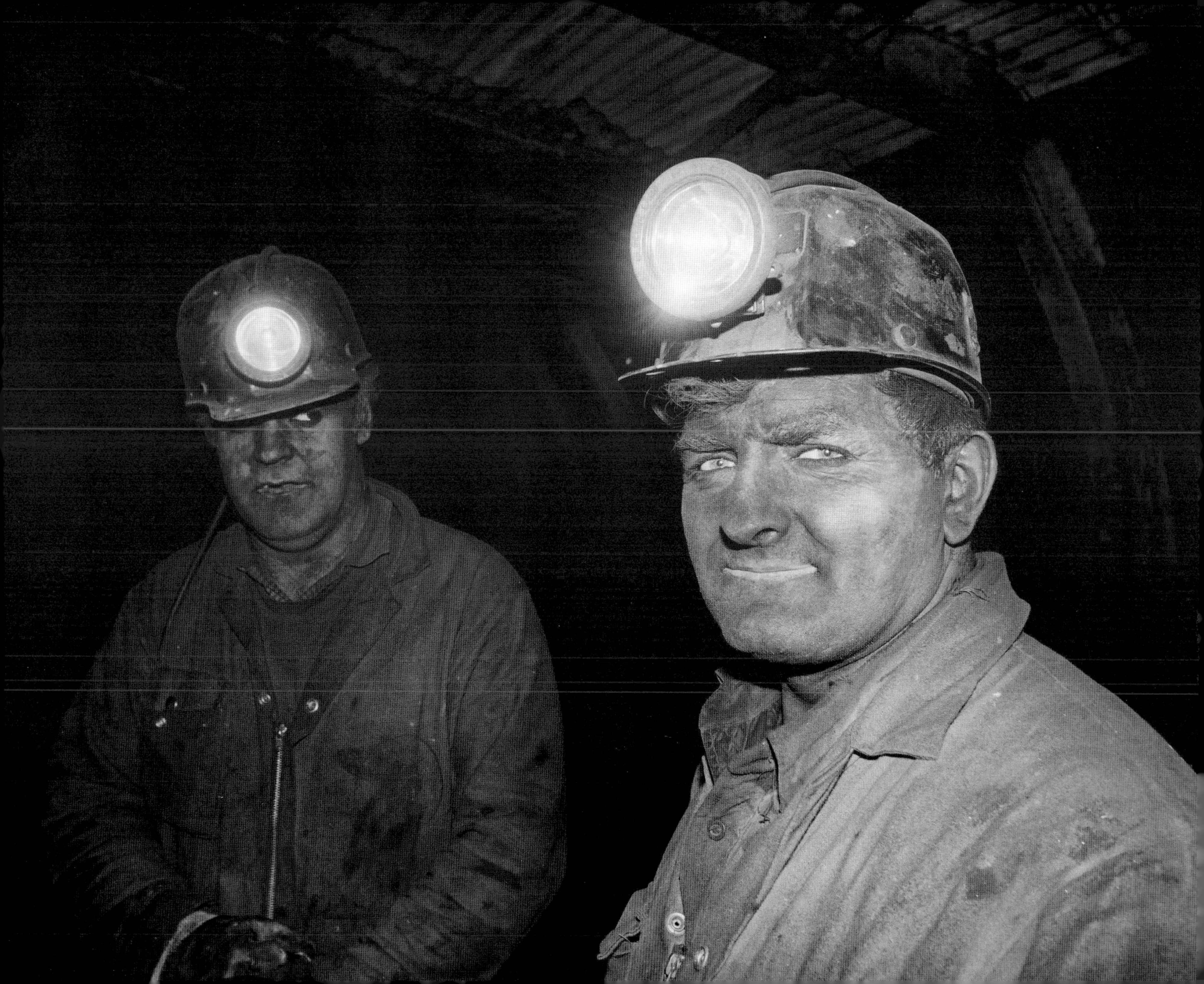

“[Mother] would have to wash his clothes separately, soaking then for almost the whole day.

Tape-2434 14 September 1984.

379-7-13A Phalen Mine 1987

Inset 379-4-35A Phalen Mine 1987

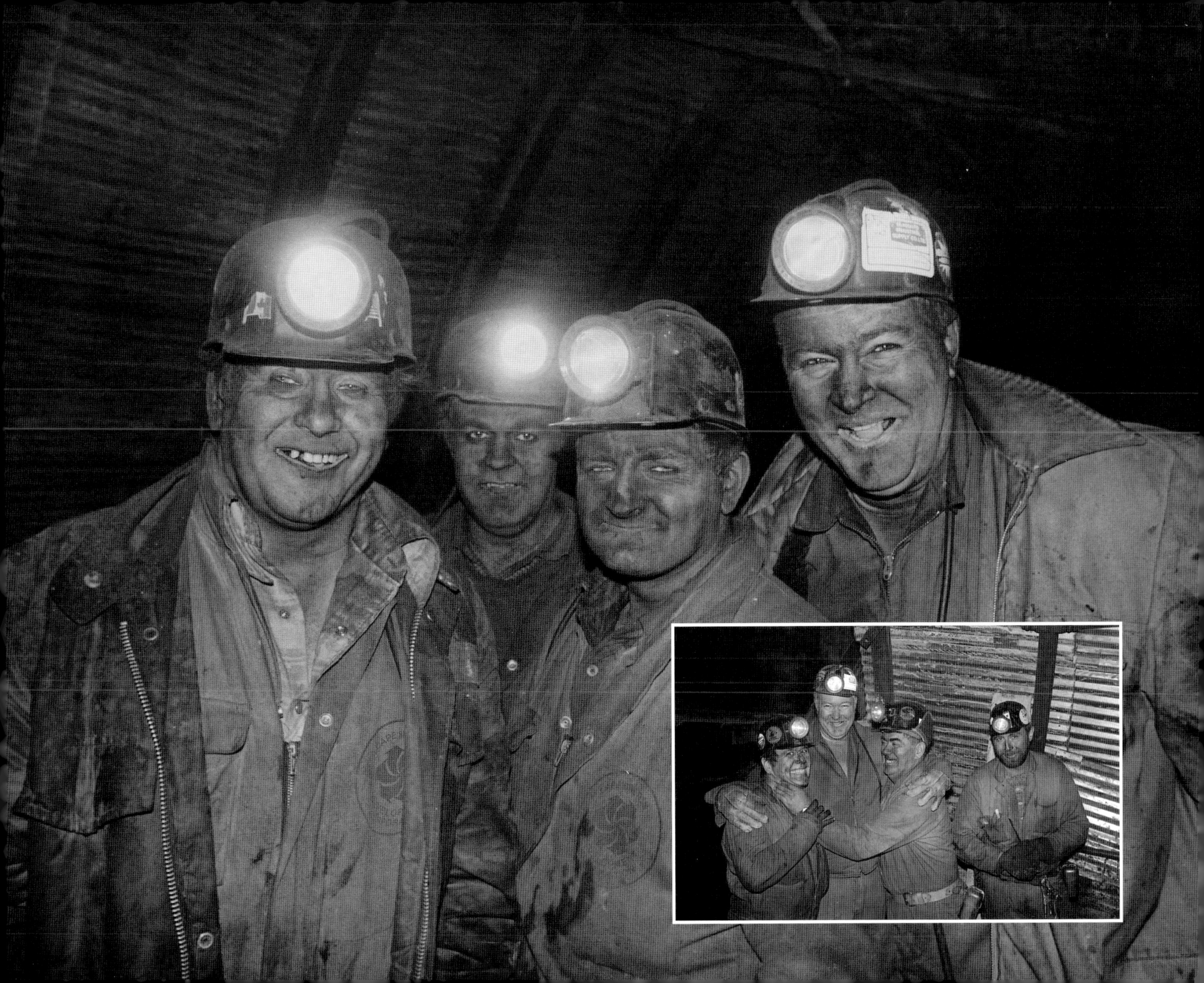

The white hats tell you they're in from the company.

In this case, it's Dowty Mining Equipment, of Britain, makers of the hydraulic pit props.

379-7-22A Phalen Mine 1987

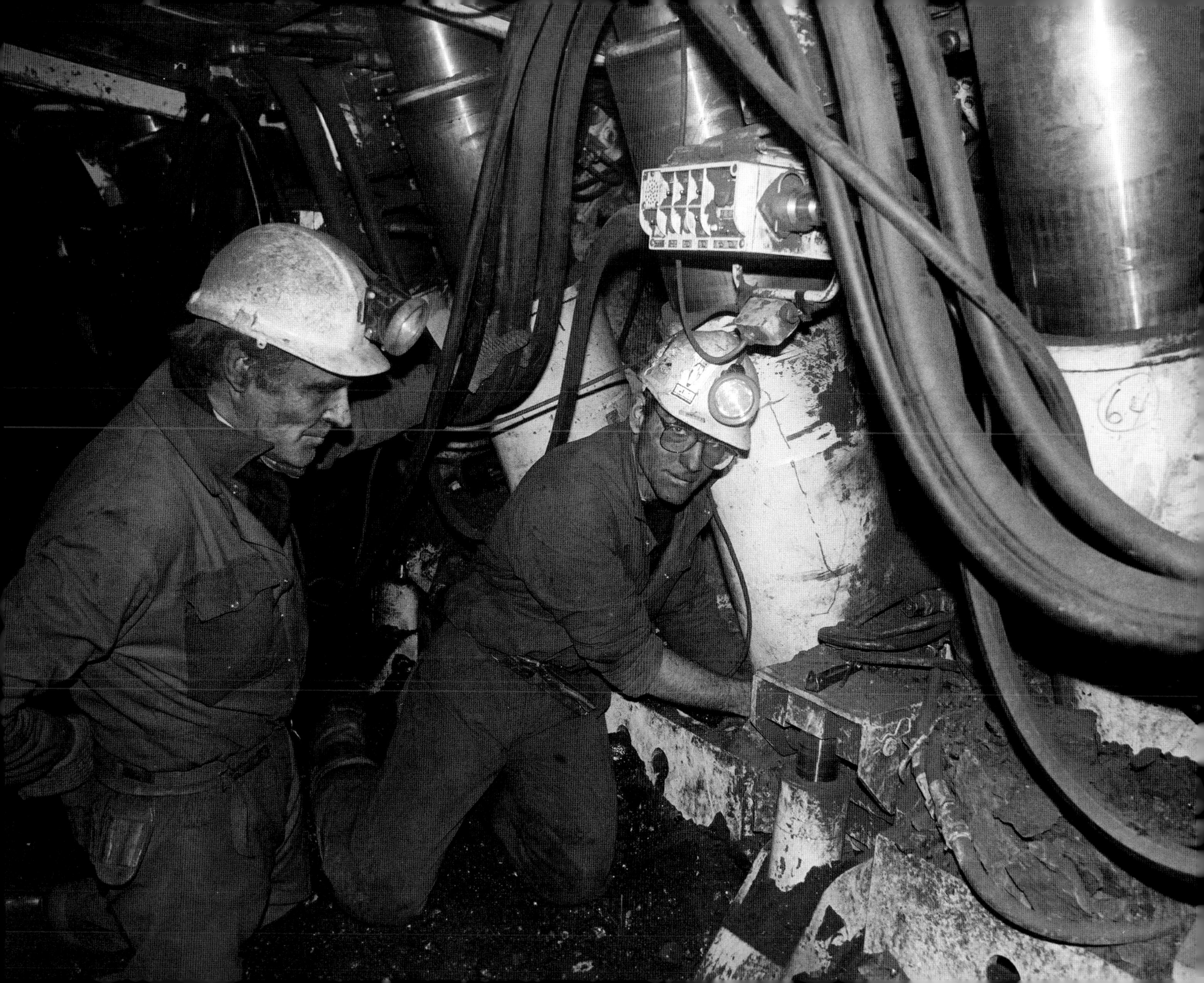
64

Phalen had the first “remote” shearer in Cape Breton.
This is the “tethered” end station.

379-7-32A Phalen Mine 1987

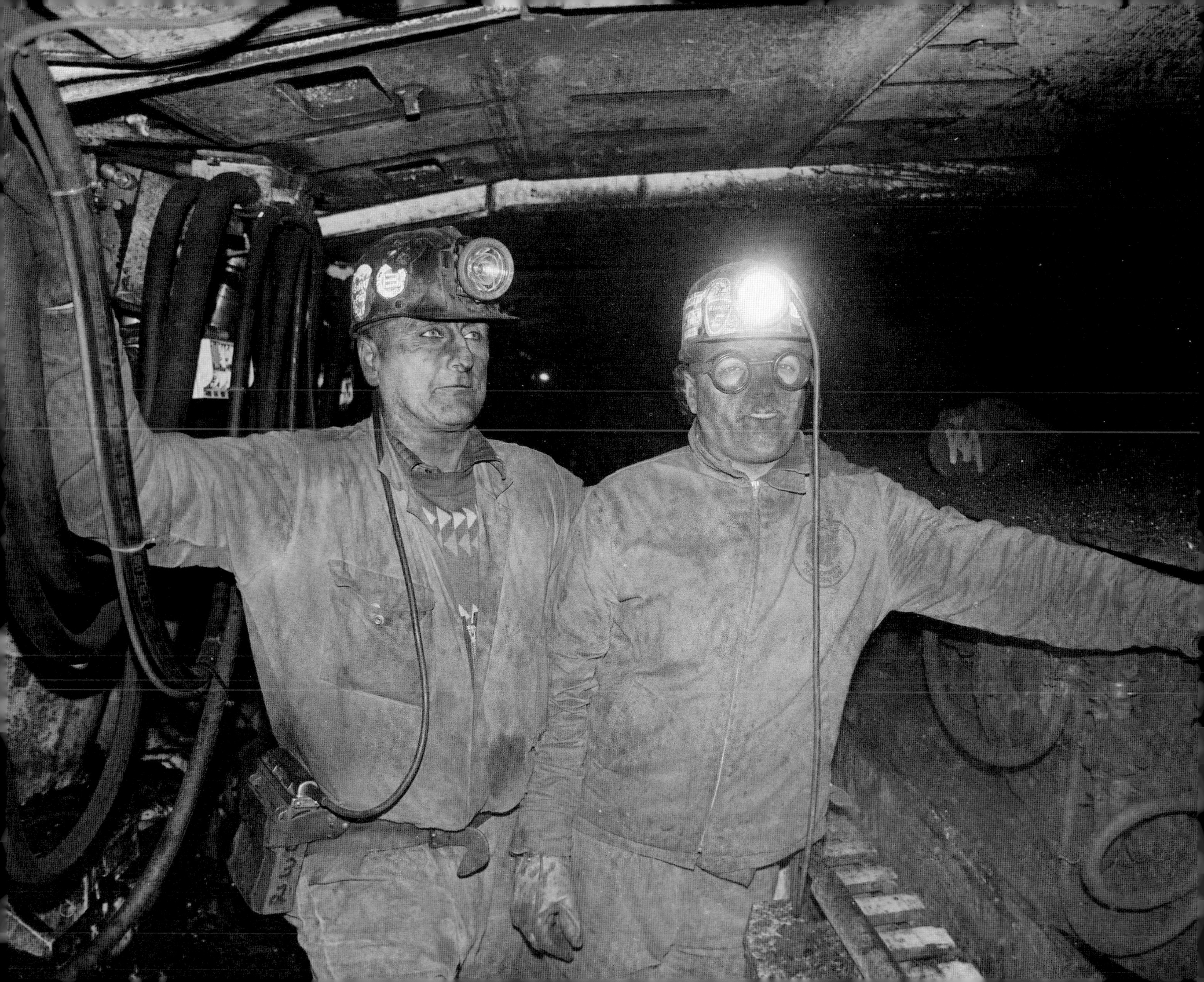

Out on the level. The console of the SIVAD
controlled the pan line and loaders.

379-5-10 Phalen Mine 1987

Inset 379-11-12A Phalen Mine 1987

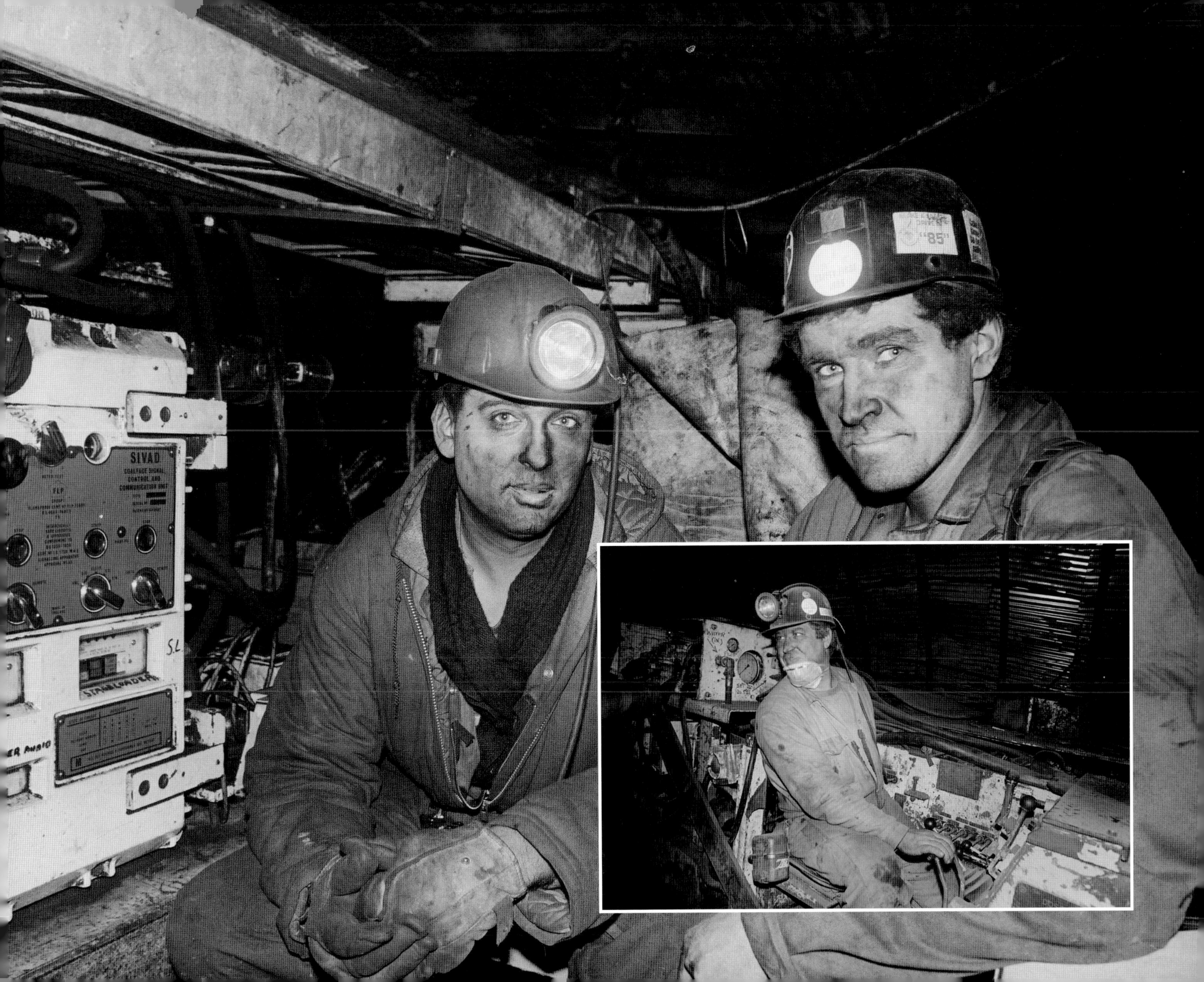
SIVAD
COALFACE SIGNAL, CONTROL AND COMMUNICATION UNIT
"85"

Phalen was so steep … they had to make an arch that was steeper on one side—a trapazoidal arch.

379-1-23A Phalen Mine 1987

Inset 379-1-22A Phalen Mine 1987

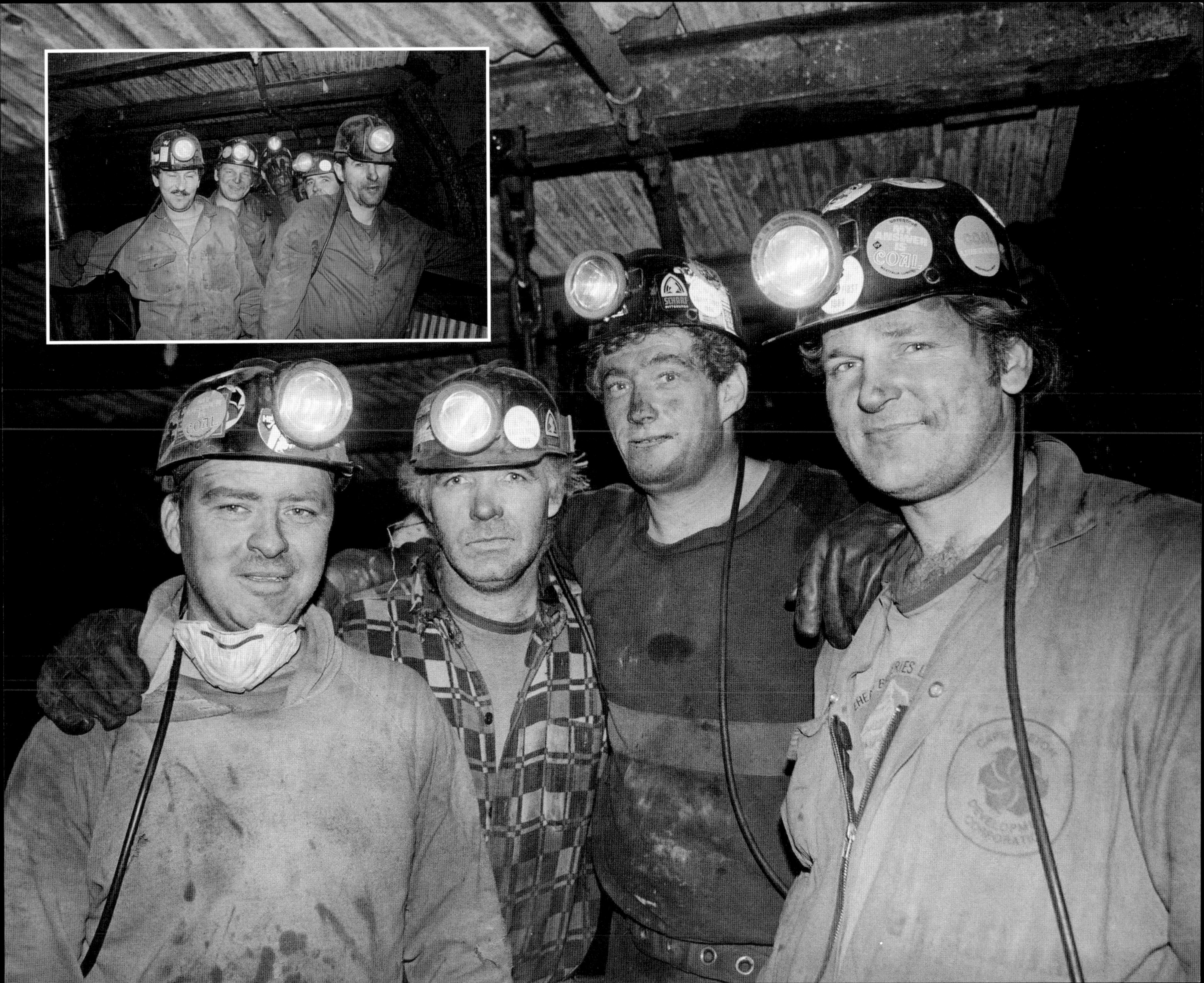

MY ANSWER IS COAL

Many men worked together forever;
they depend on each other and become good friends.

See the stickers on their helmets? Suppliers used to bring stickers. Some guys had stickers all overtheir helmets—they were handy for identifying your lunch can, too.

Clockwise from top left
380-E-17A Phalen Mine 1988

379-4-3A Phalen Mine 1987

379-9-24 Phalen Mine 1987

379-4-20A Phalen Mine 1987

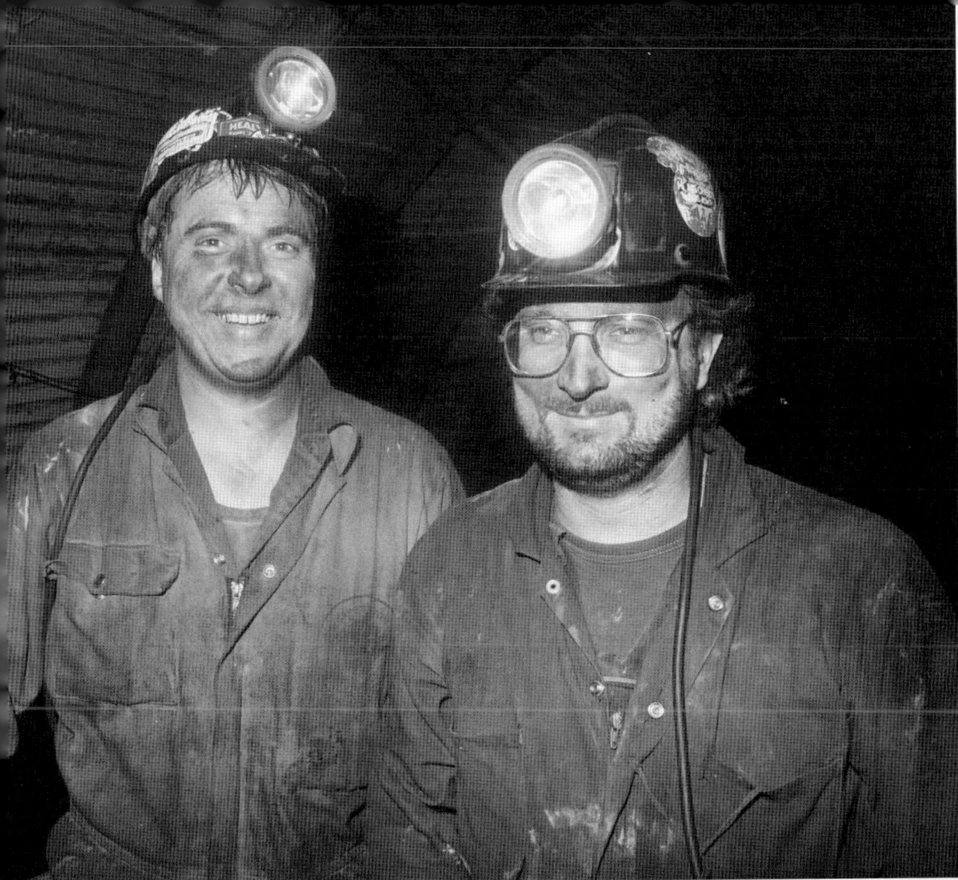

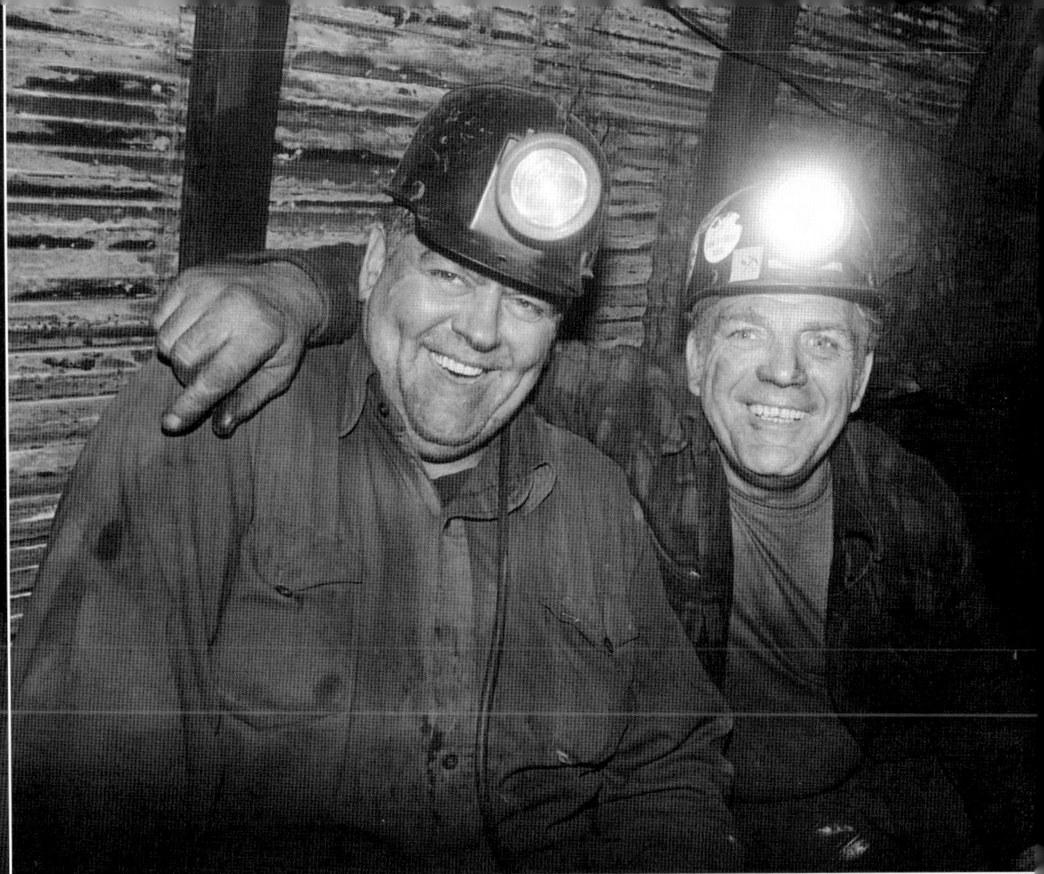

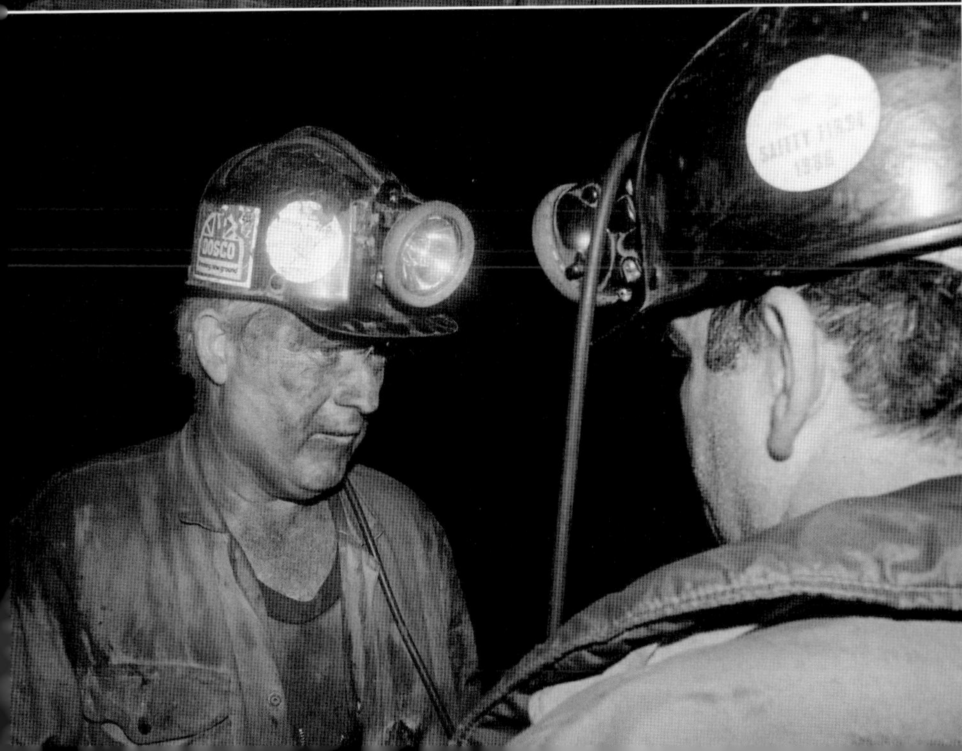
DOSCO
SAFETY FIRST

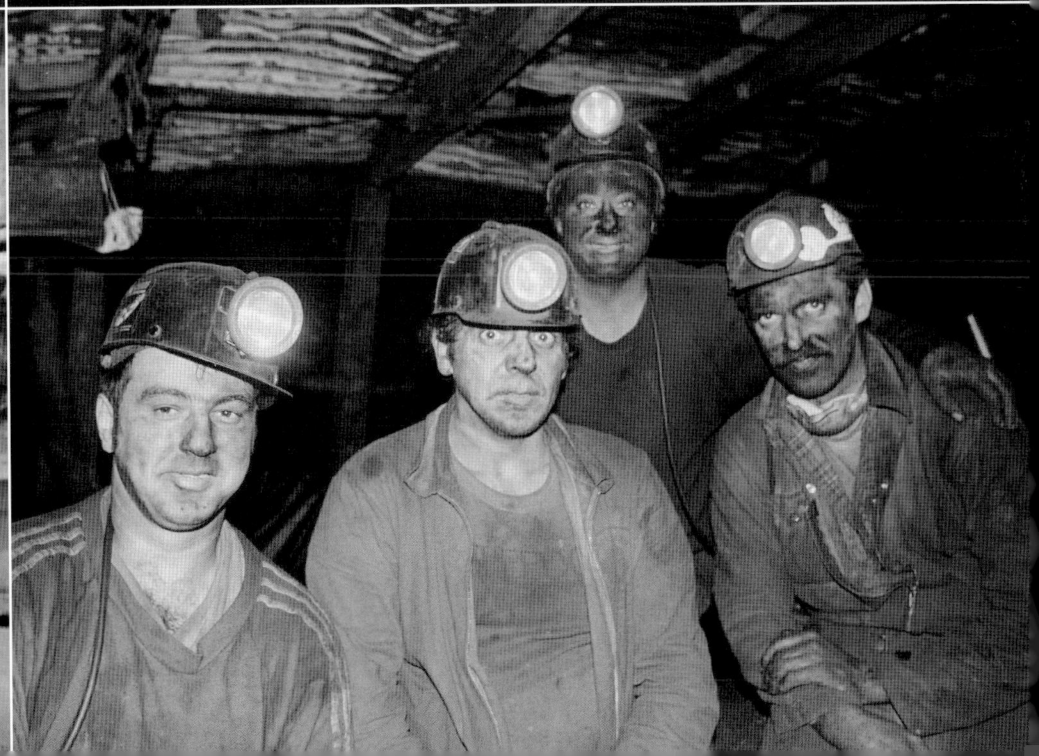

With the dirt of our hands you'd eat around the sandwich except for the part you were holding, we threw that away because of the dirt, feed it to the rats.

You ate when you first went in then work for 8 hours because there was no set lunch times and we were paid based on production.

The speaker on the wall is what we used for communications.

379-9-35 Phalen Mine 1987

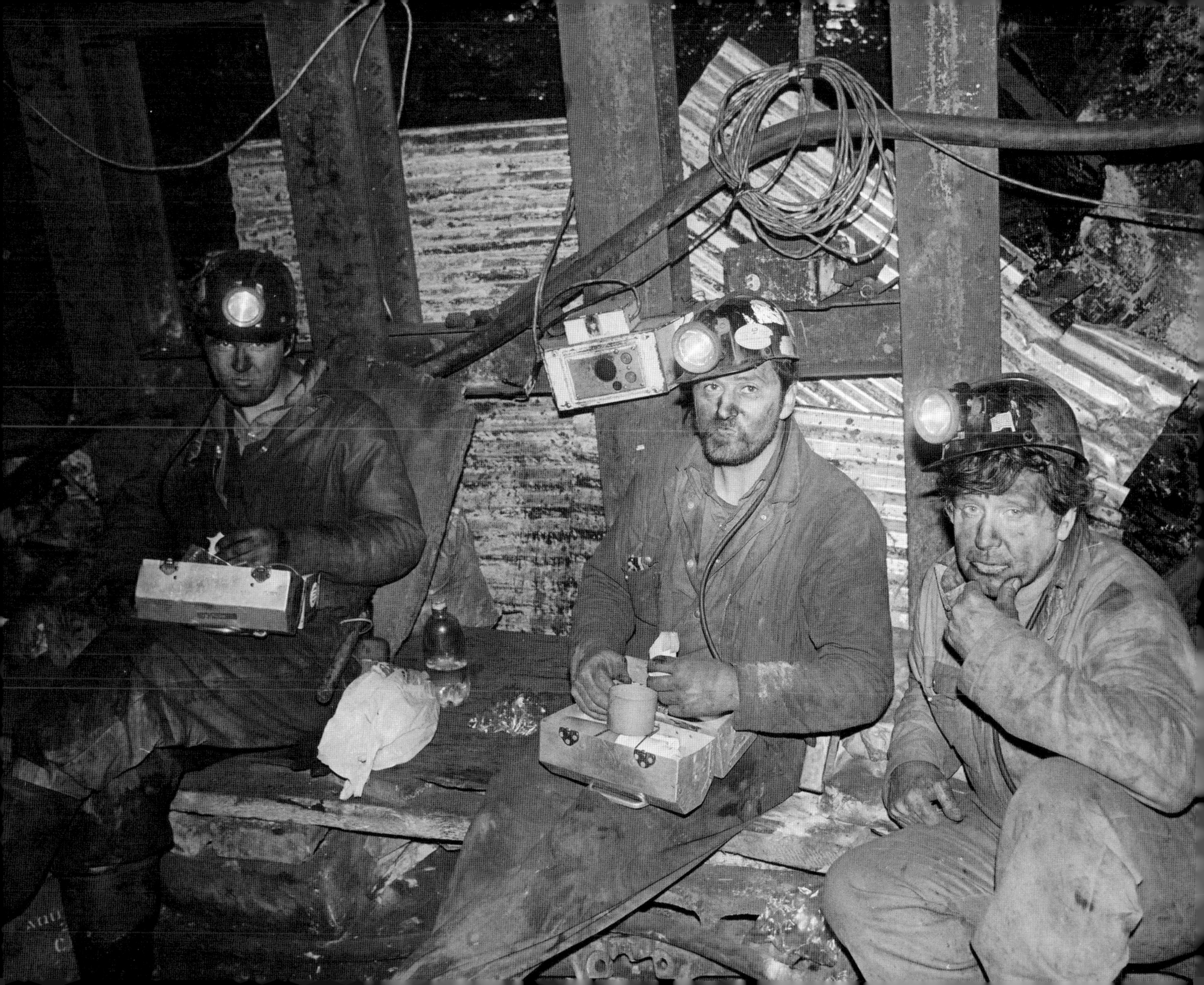

Driving the deep, the slope. Supporting the roof of the main tunnel using steel arches.

Note the swirls in the roof from the picks of the roadheader.

379-6-27A Phalen Mine 1987

Left Inset 379-6-29A Phalen Mine 1987

Right Inset 379-6-24A Phalen Mine 1987

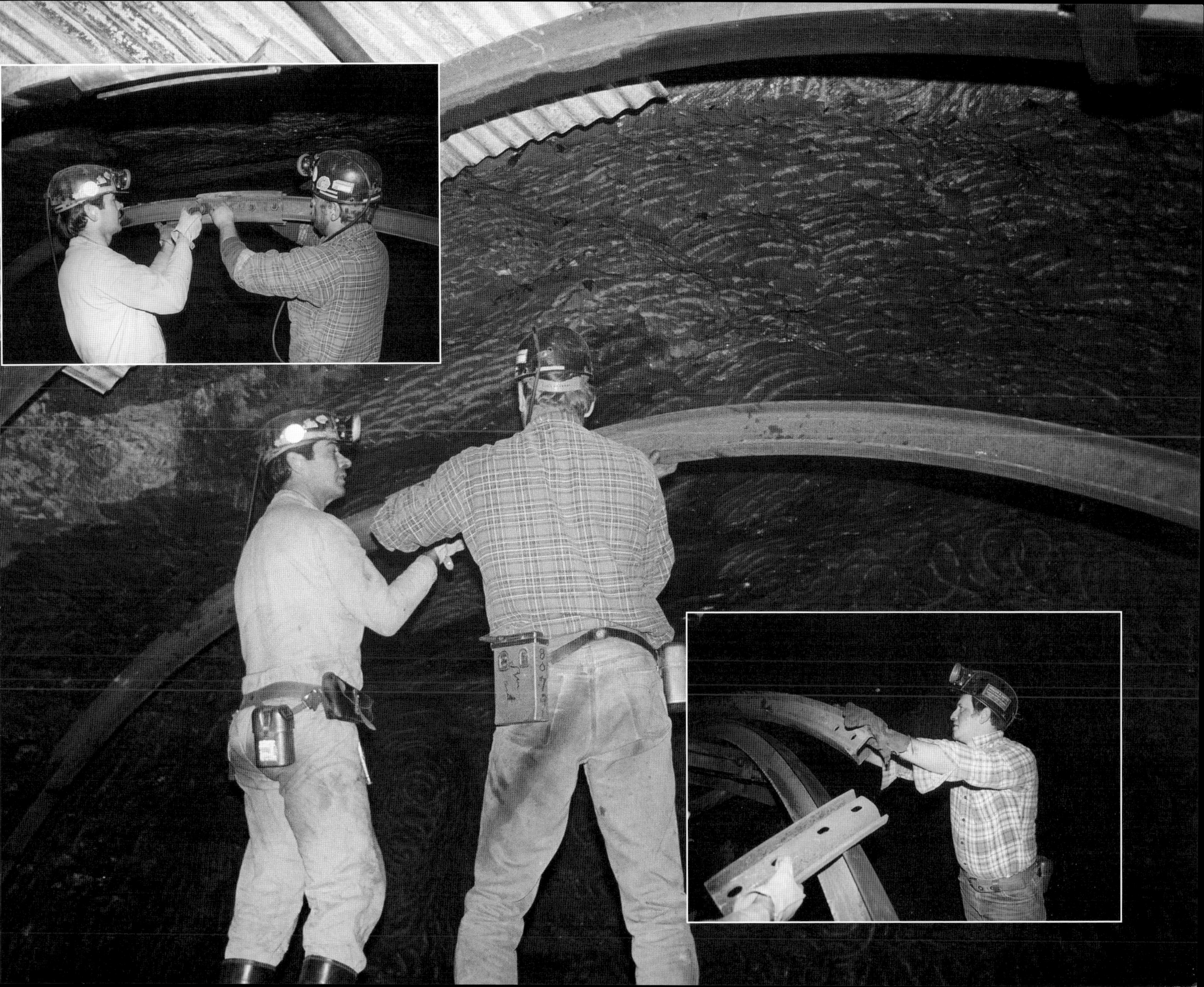

Steel boomers. Putting in the sheeting. Must be an intersection—notice the wood blocks.

Clockwise from top left
380-D-20A Phalen Mine 1988

380-D-15A Phalen Mine 1988

380-D-10A Phalen Mine 1988

380-D-19A Phalen Mine 1988

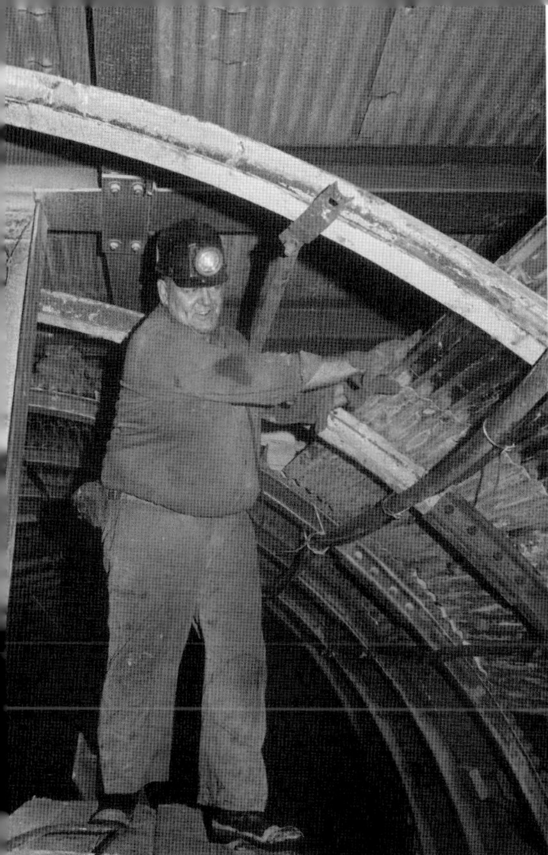
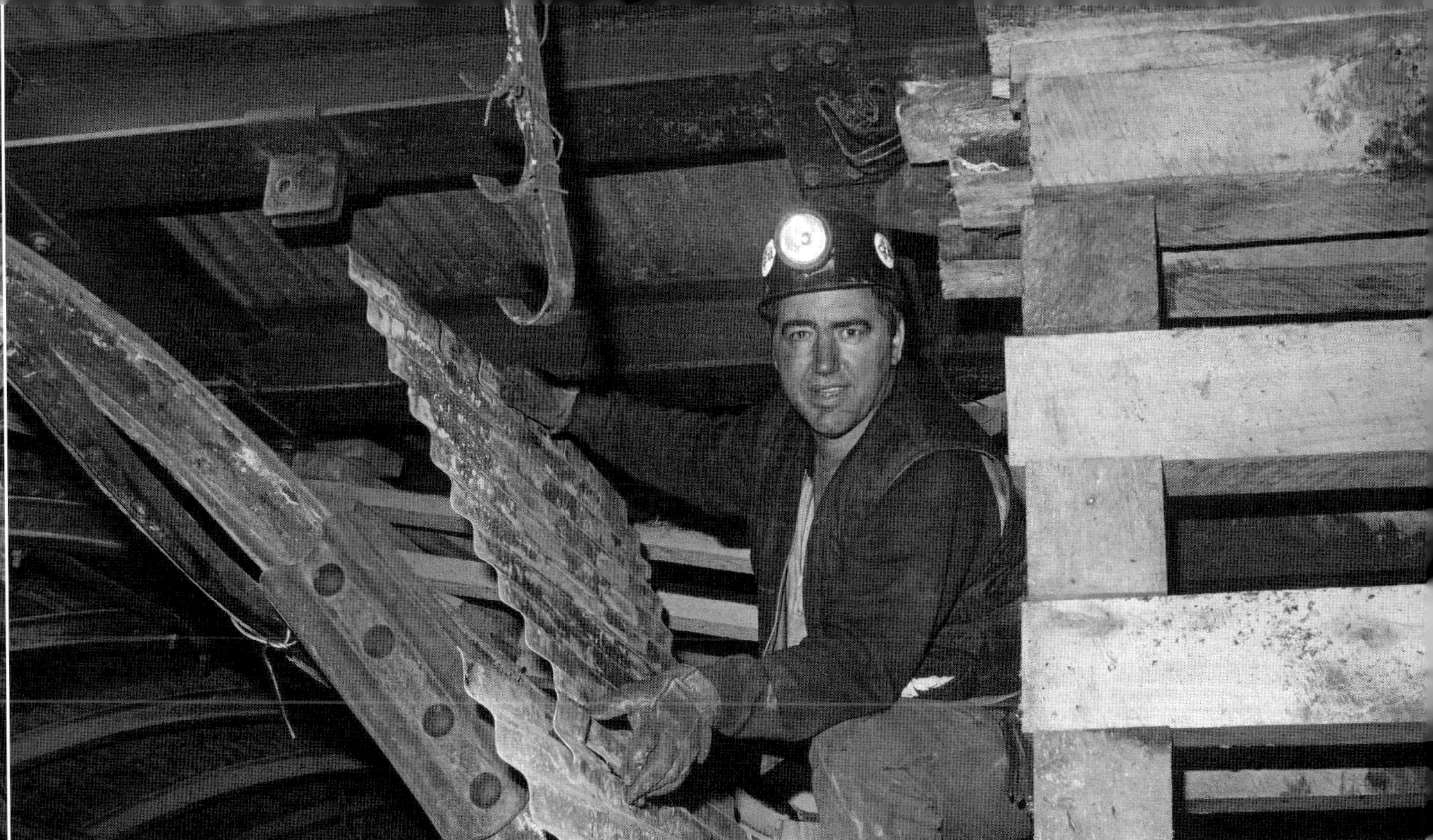
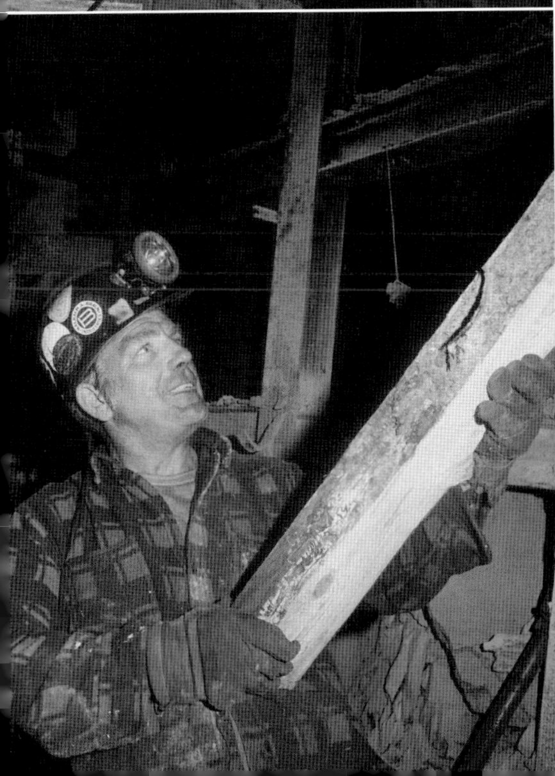
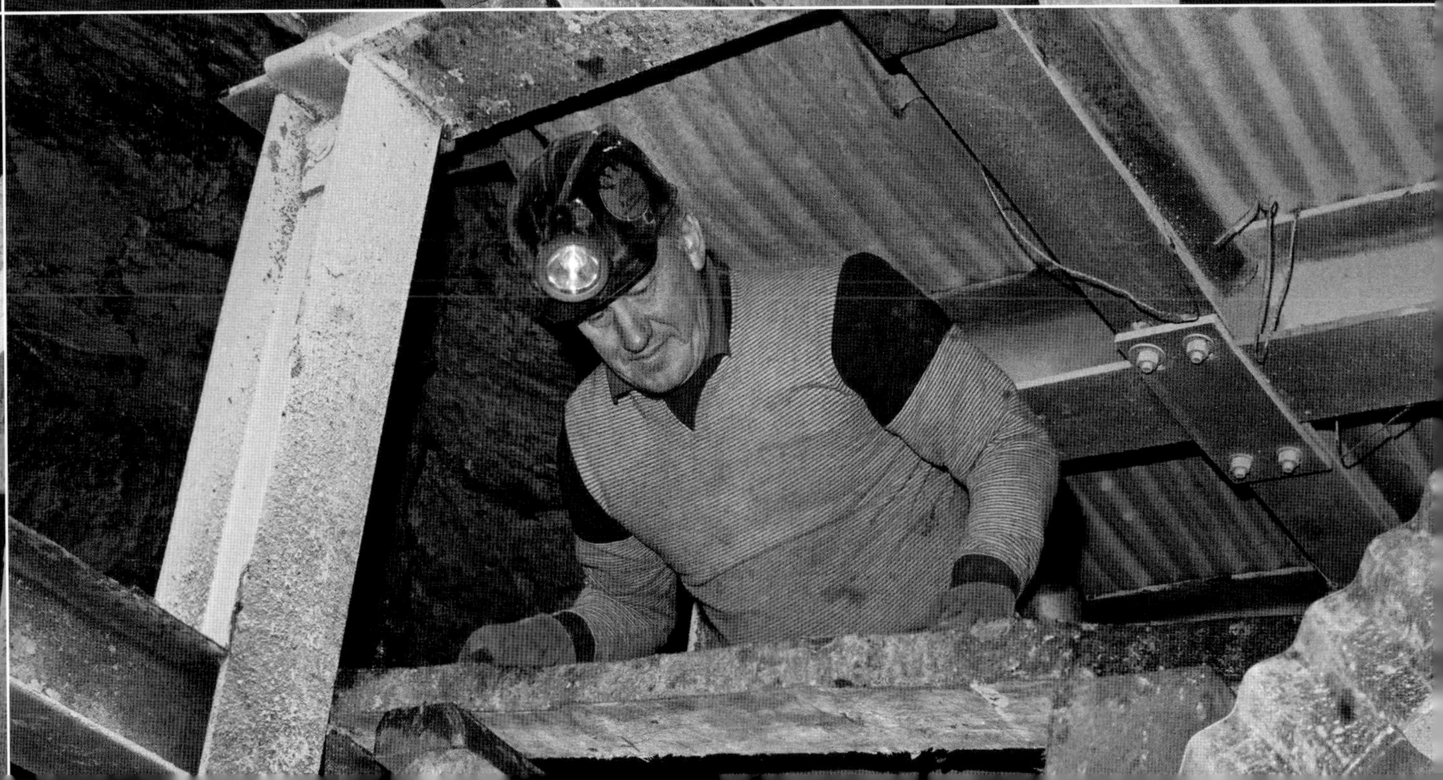

The monorail, suspended from the 16 ft ceiling, was used to move equipment through the mine.

379-9-34 Phalen Mine 1987

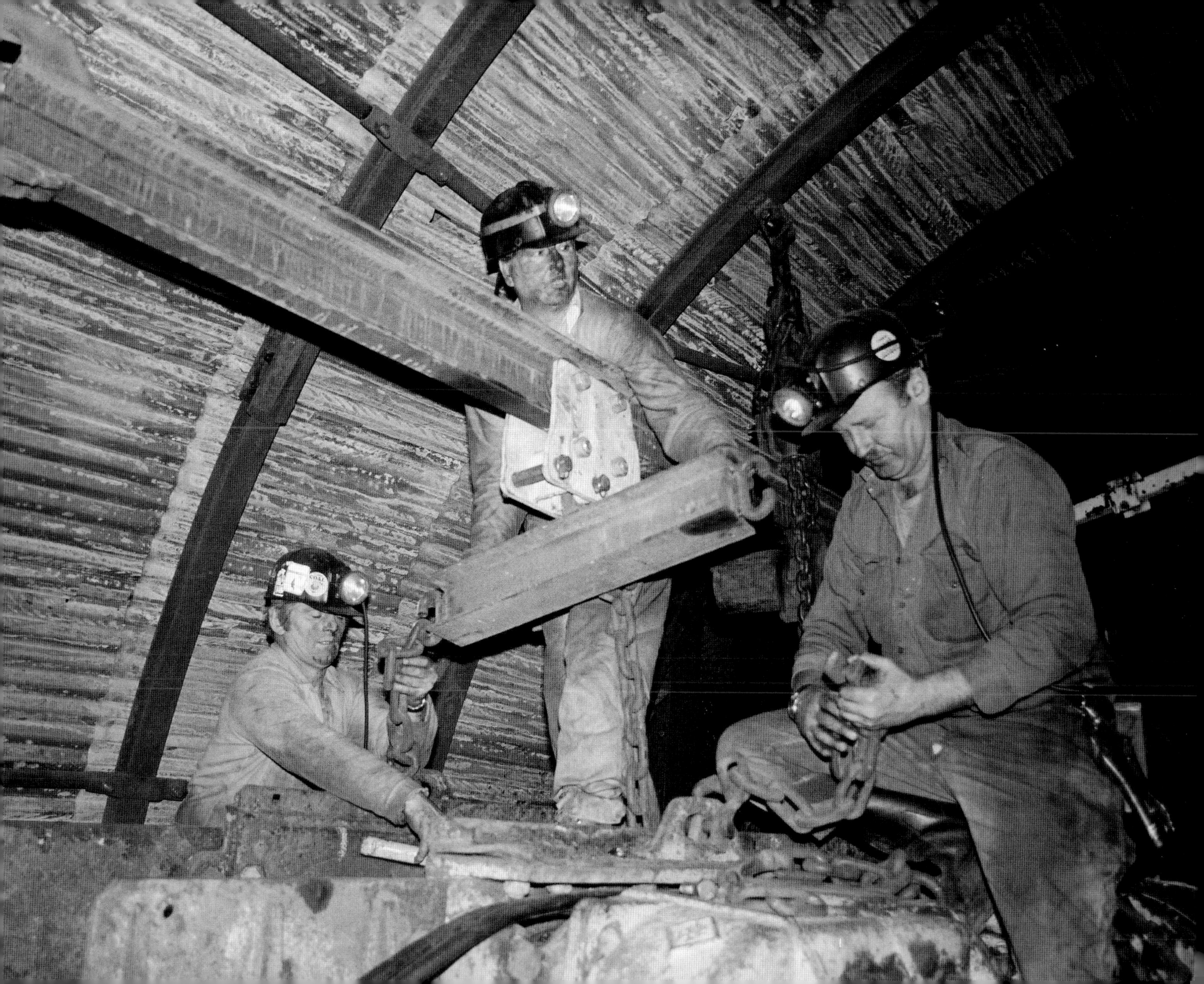

Note the number written on the jack post. Each post was numbered—a jack post every 3 ft (approx. 1 m). This post was 216 ft (approx. 65 m) along.

380-A-31 Phalen Mine 1988

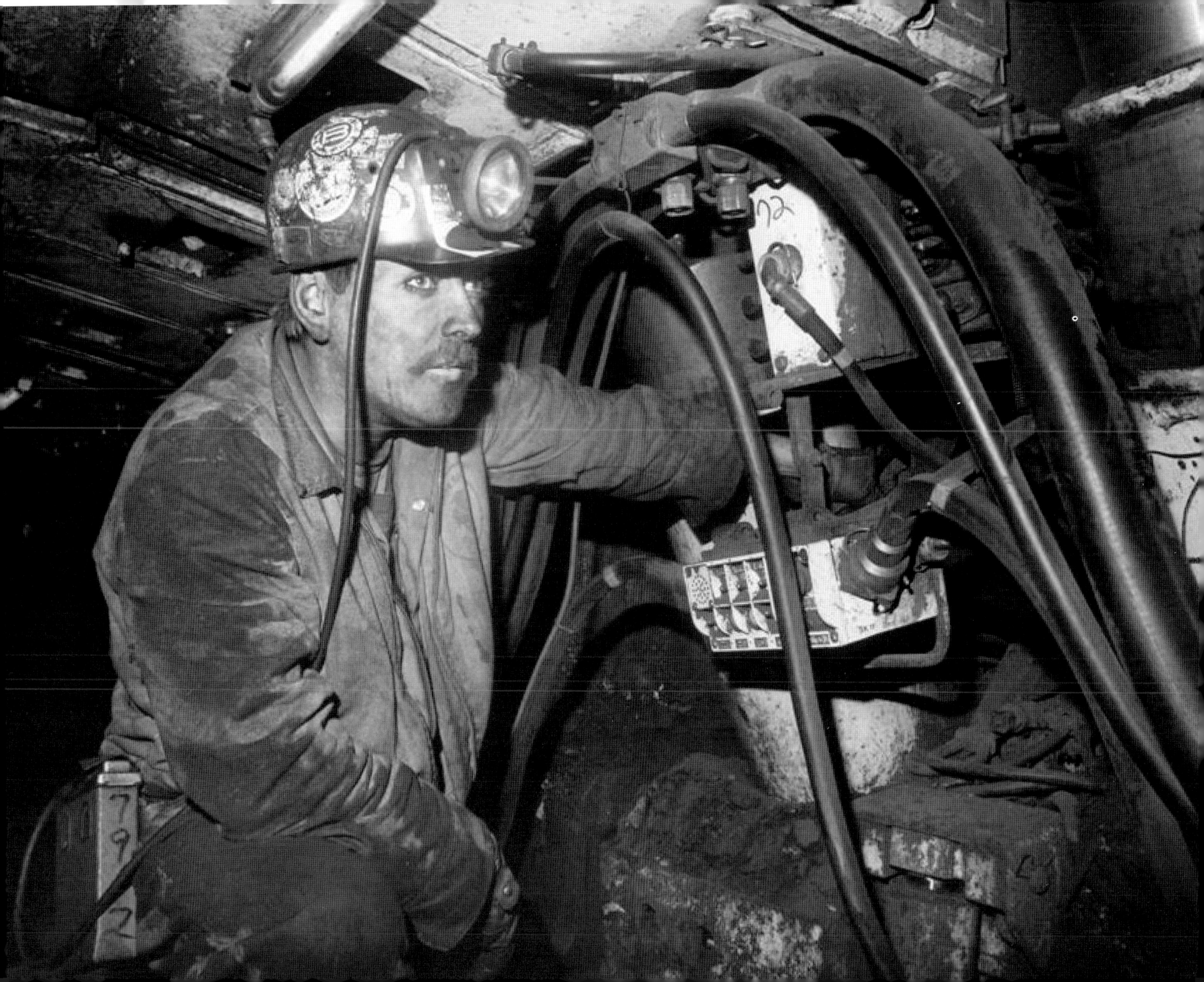

The “bag” above their hand that’s an air bag,
where the air came in (ventilation).

Men laying rail, bags of stone dust,
water line and fire hydrant on the left.

379-13-34A Phalen Mine 1987

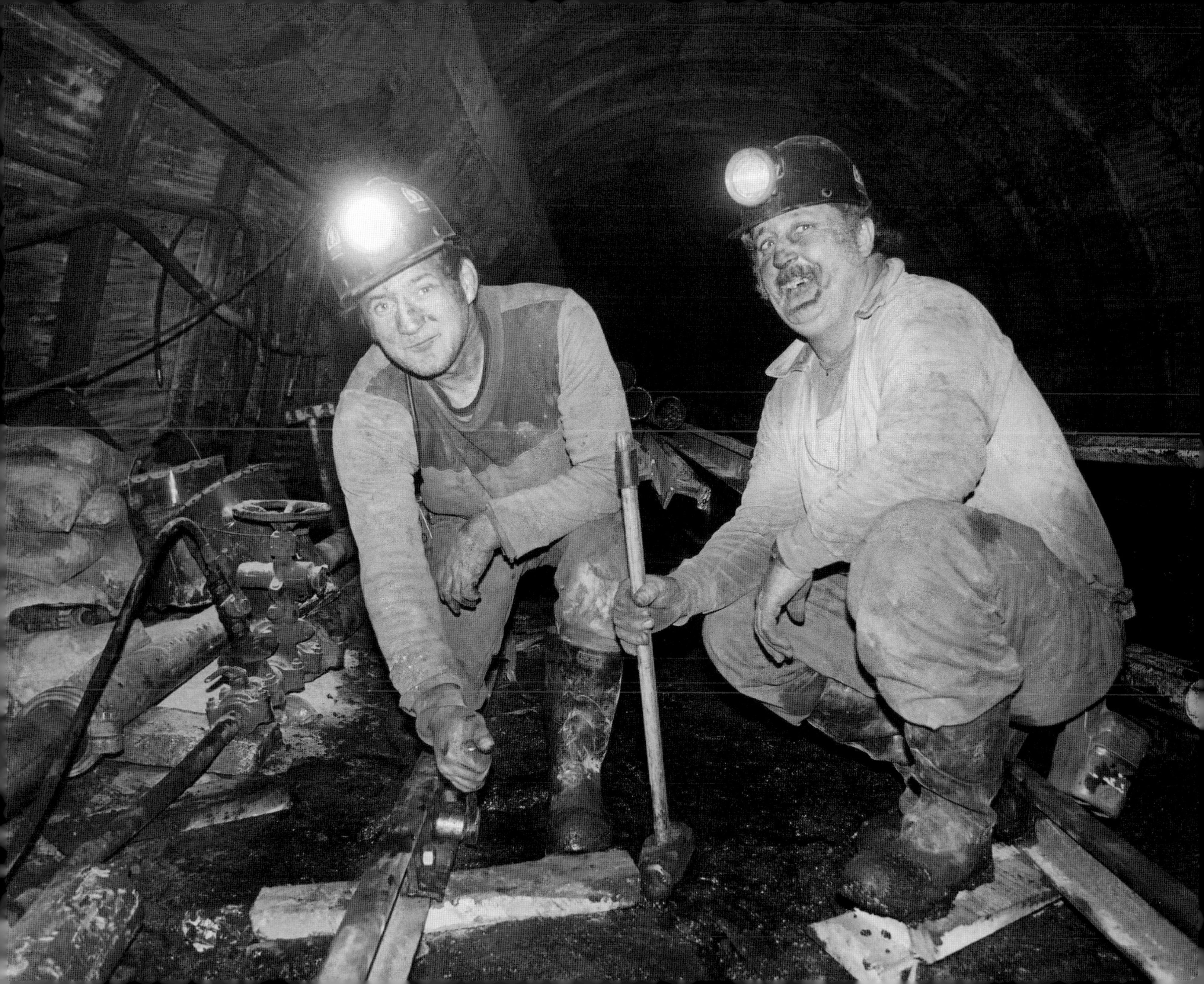

White helmet, means he's from outside, an inspector.
Inspectors could shut everything down.

380-A-23 Phalen Mine 1988

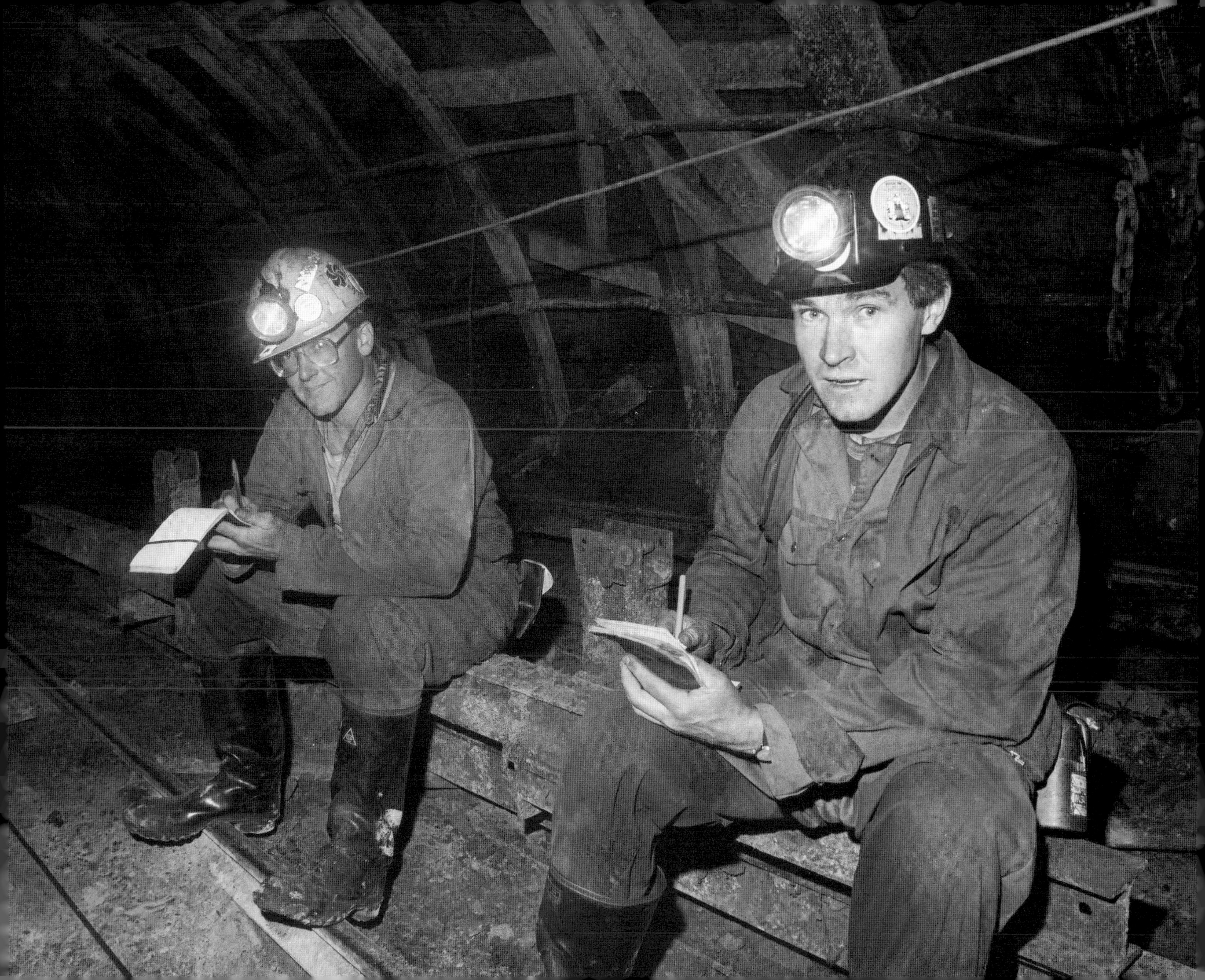

Stoneduster.

They monitored the amount of coal dust, which was flammable; if it got too high, work shut down and they applied the stone dust—ground limestone. Renders coal dust inert.

380-C-12 Phalen Mine 1988

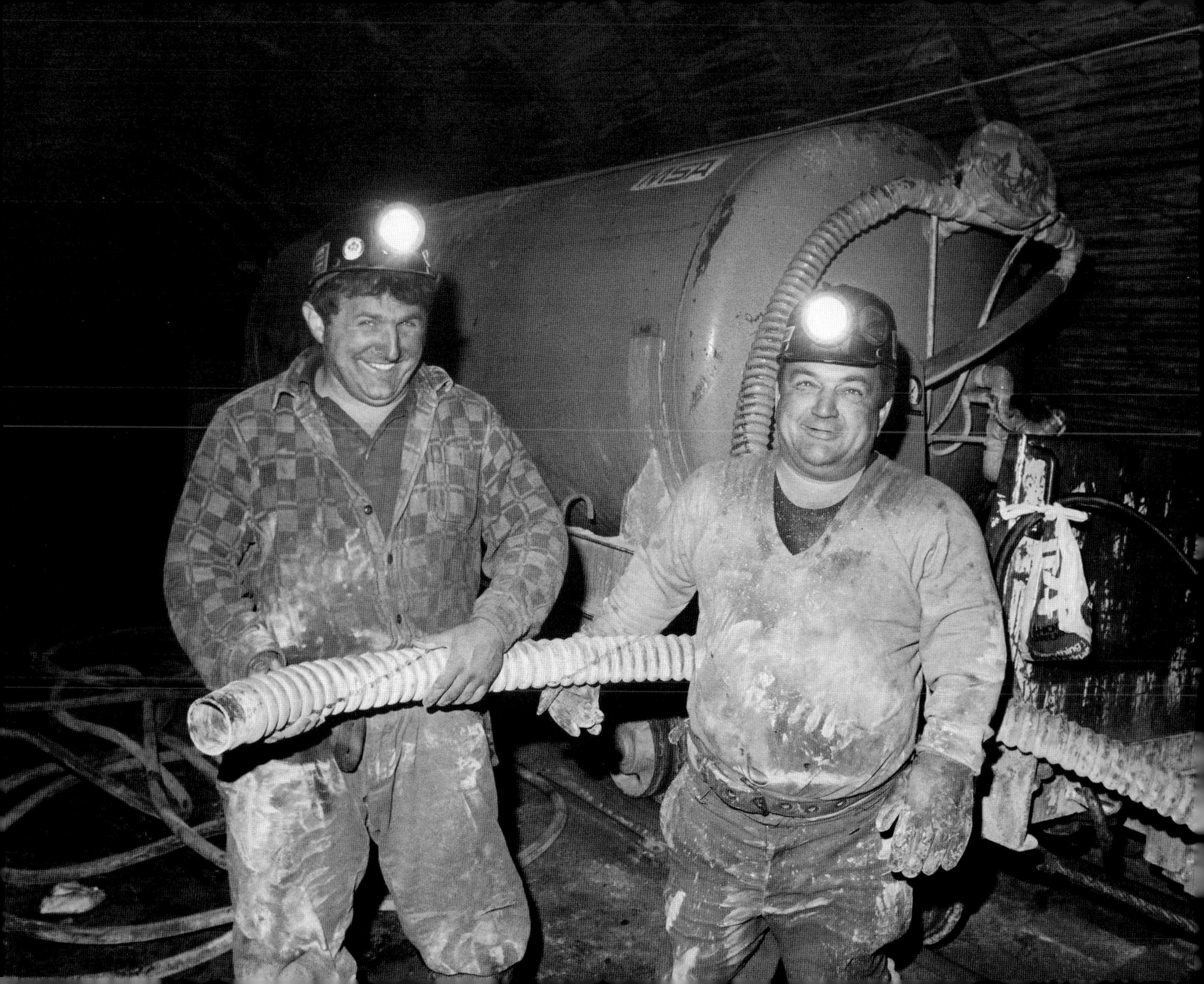

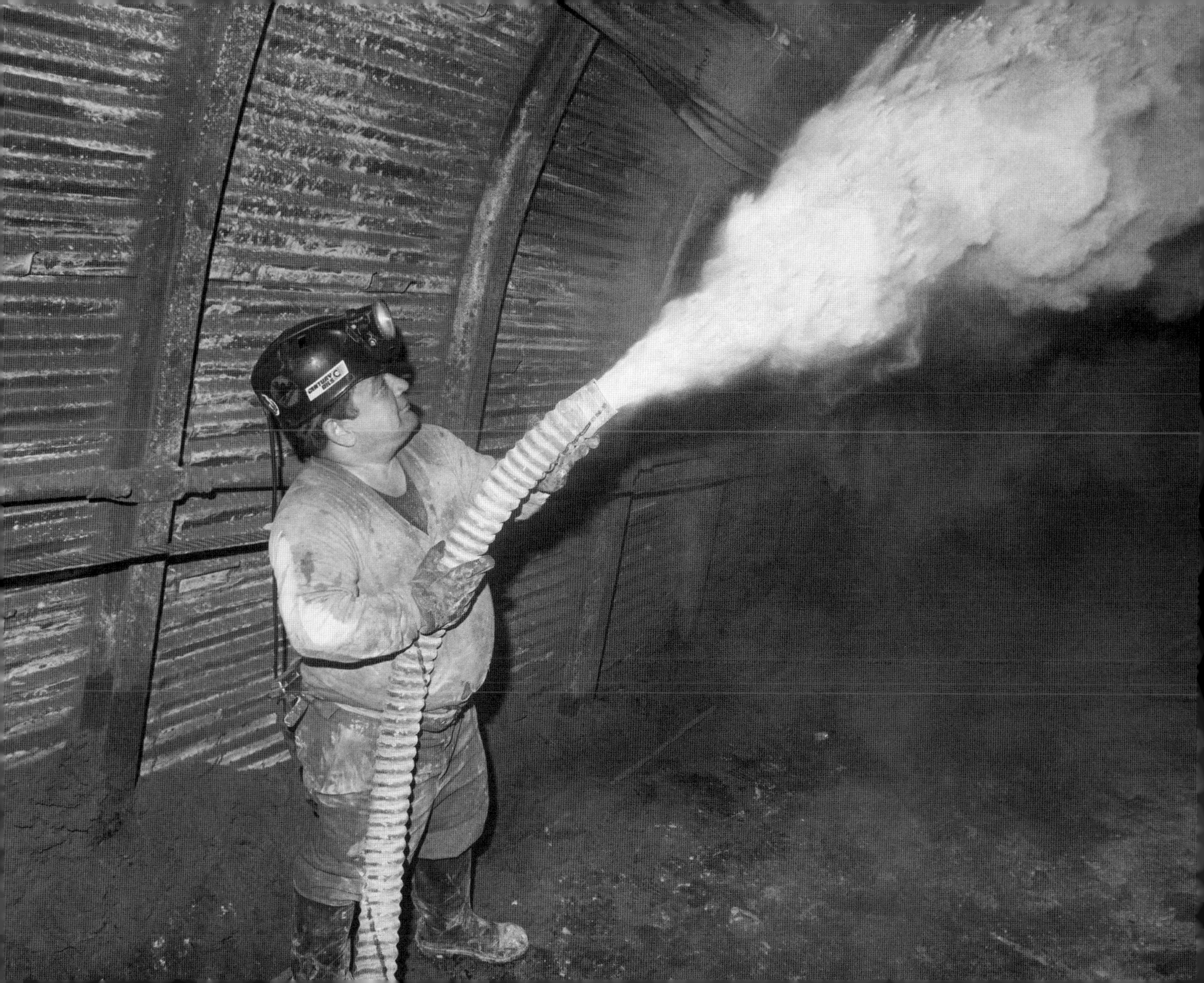
CENTURY

380-B-11A Phalen Mine 1988

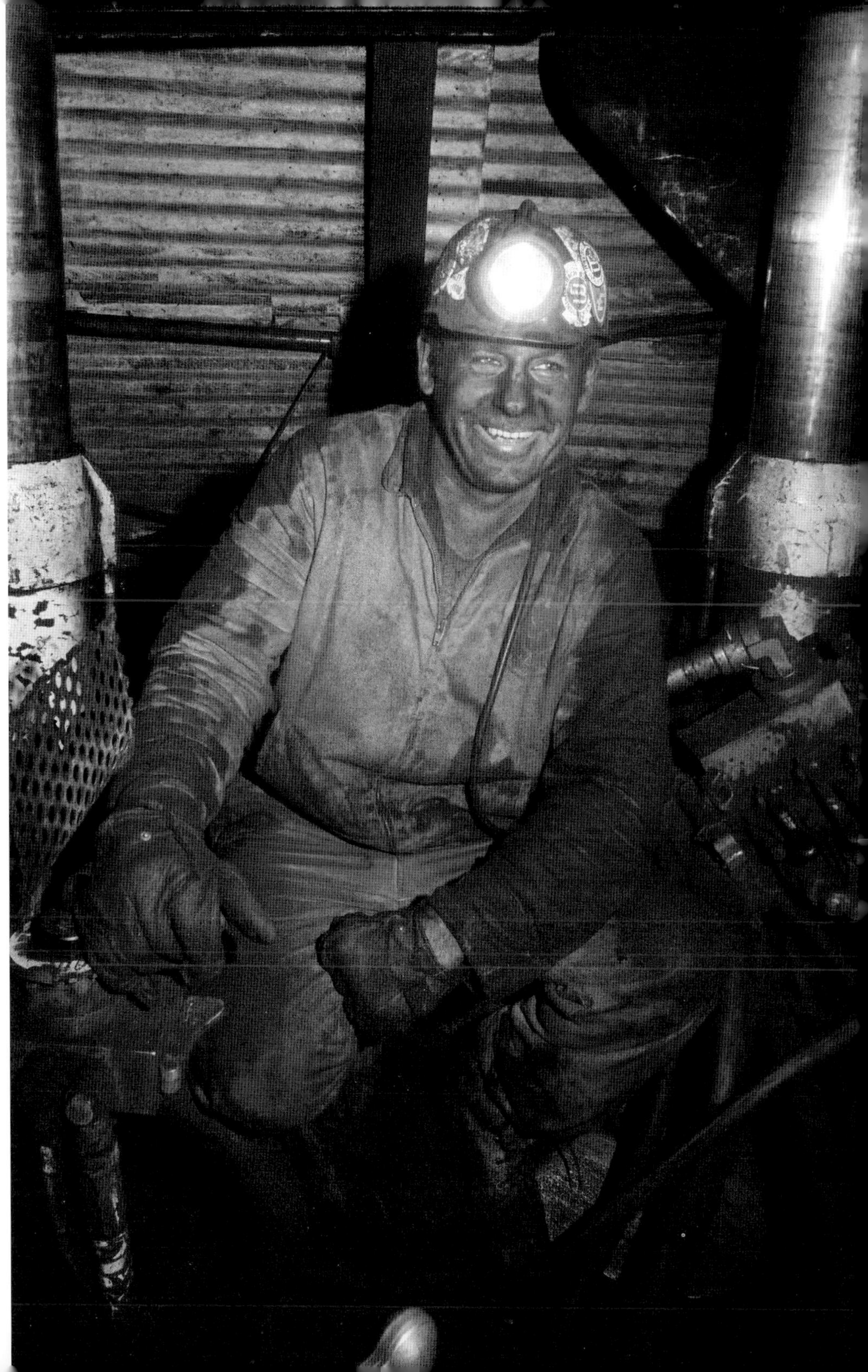

144-A-10 No. 26 Colliery, 1981

No. 26 Colliery
1975
1977
1981
1983
1985

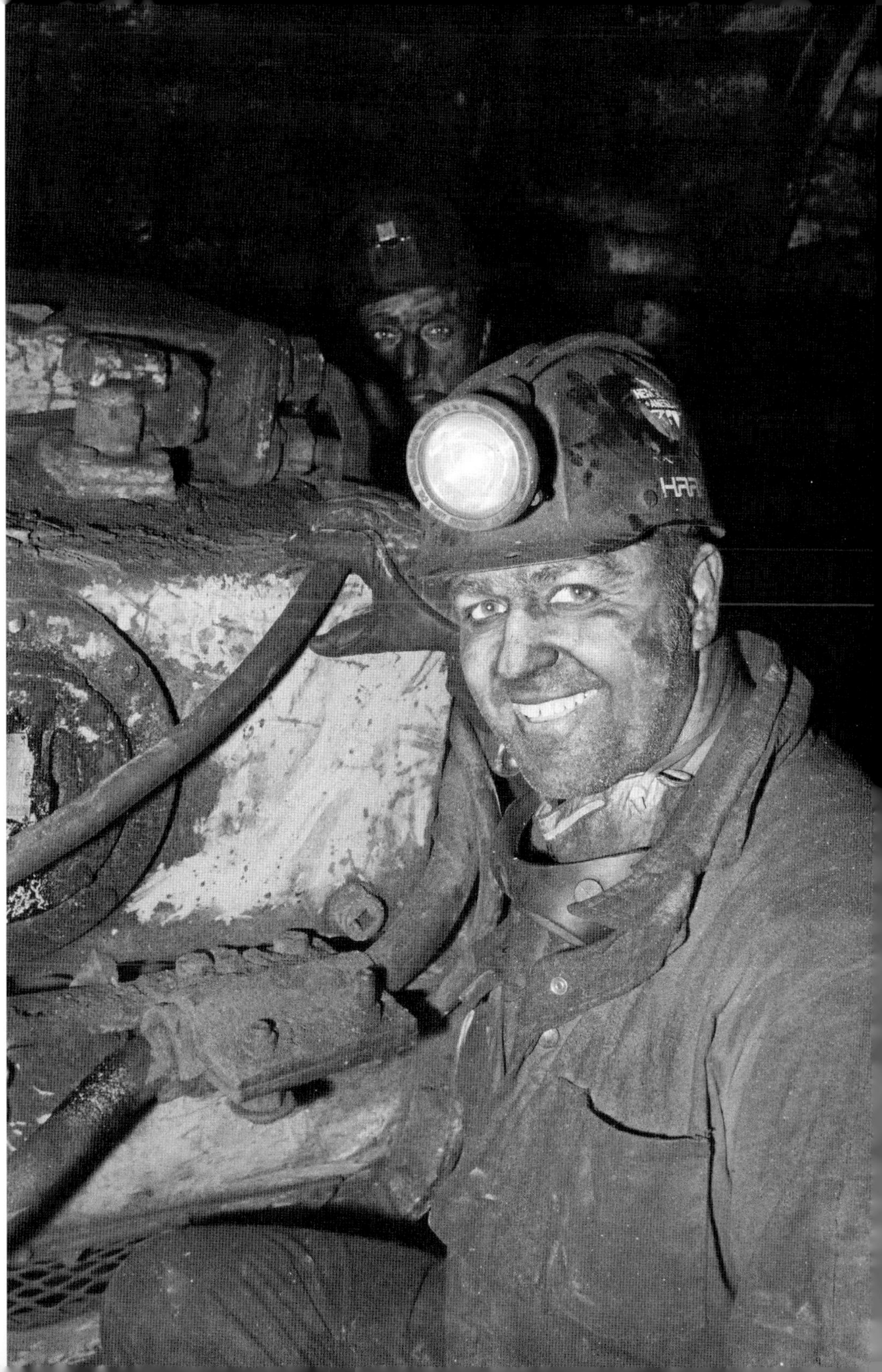

In No. 26, the shearer moved along the wall face on a haulage chain. The chain was under very high tension.

144-A-19 No. 26 Colliery 1981

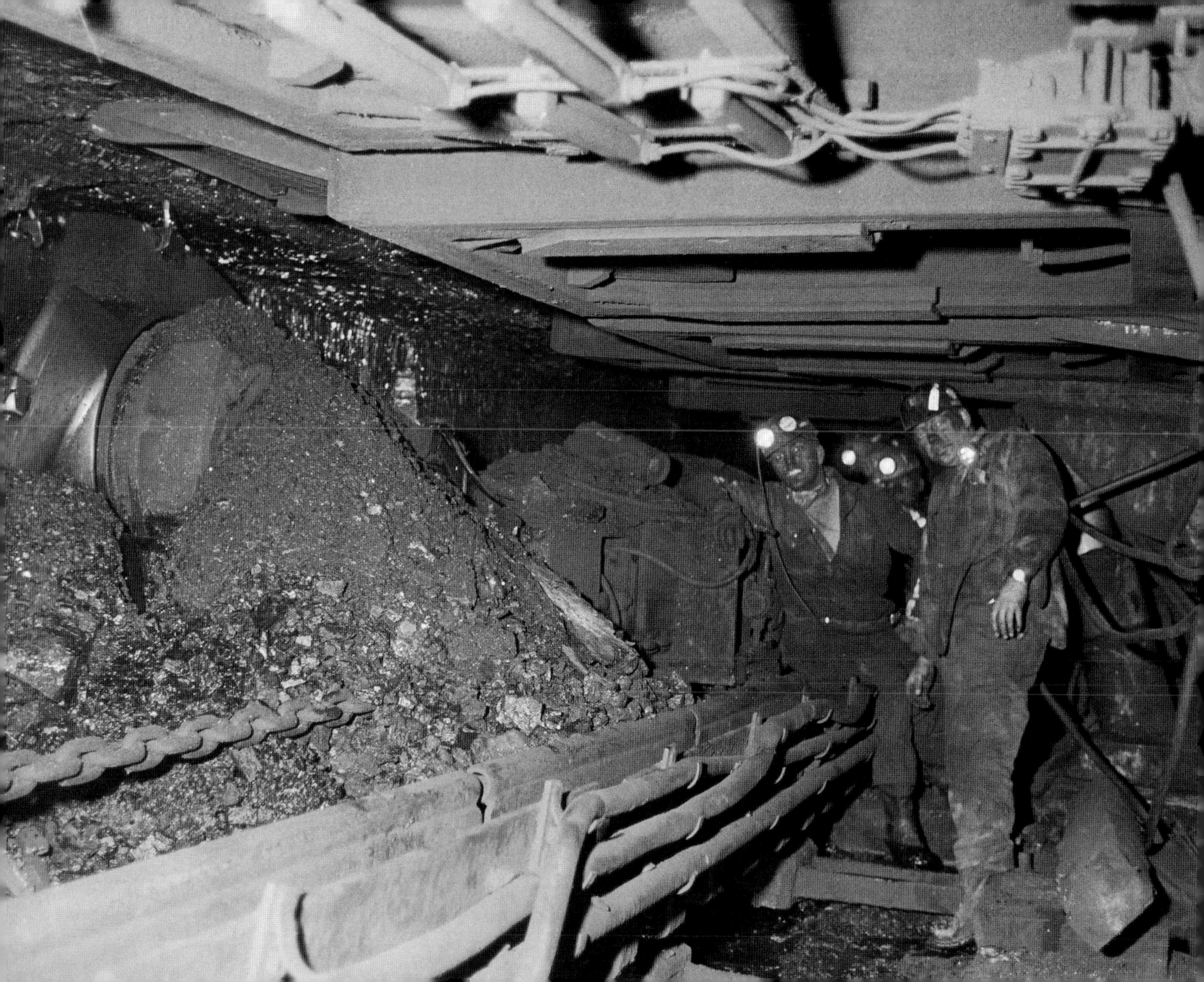

No. 26 used Heimscheidt roof supports. Note the loose rock above the support.

144-A-17 No. 26 Colliery 1981

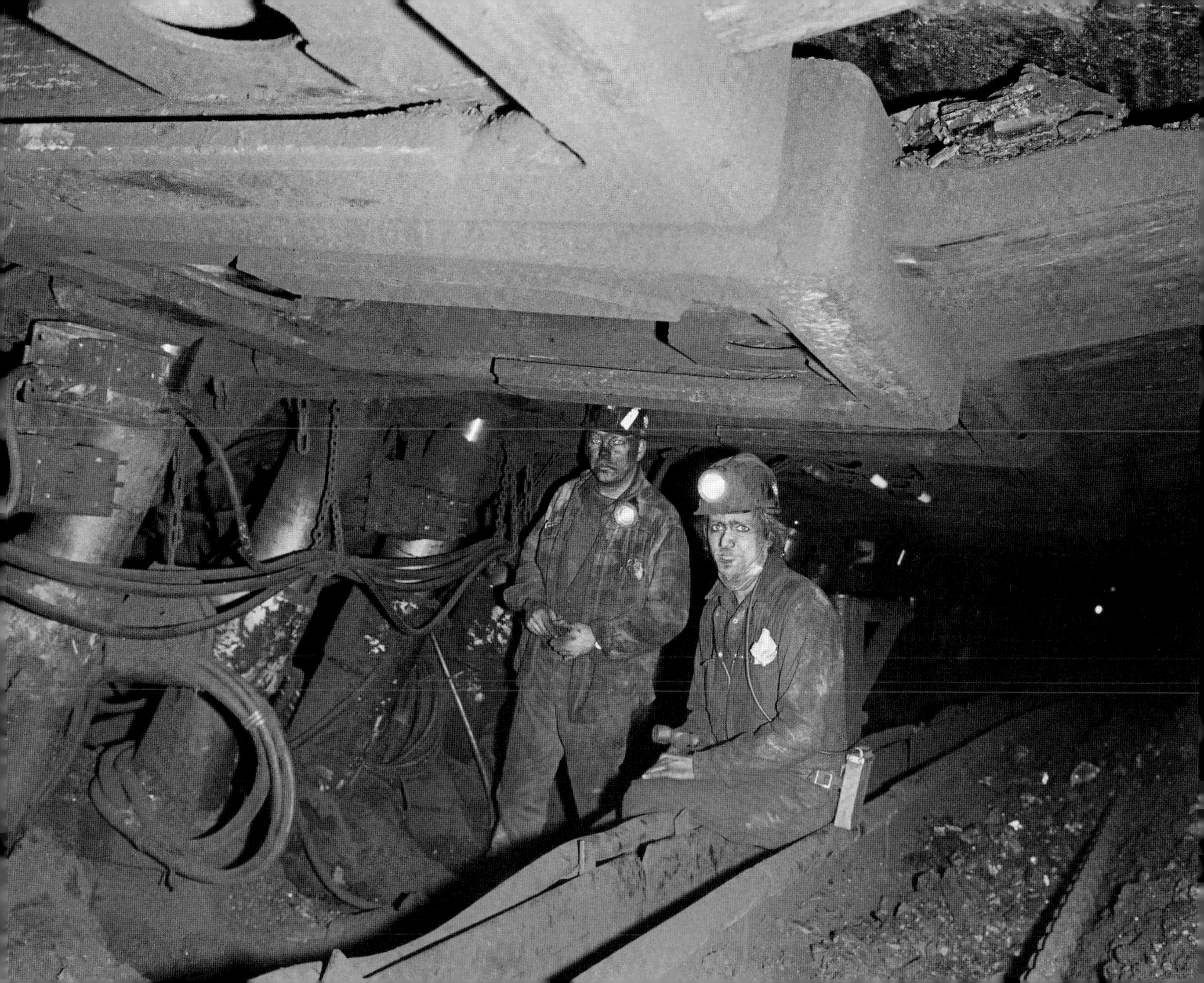

Operating an EIMCO. Notice the water bottle, there was no water down there you had to bring in all that you needed. Everywhere you can see the water bottles hanging on hooks, nails ... everywhere.

144-A-11 No. 26 Colliery 1981

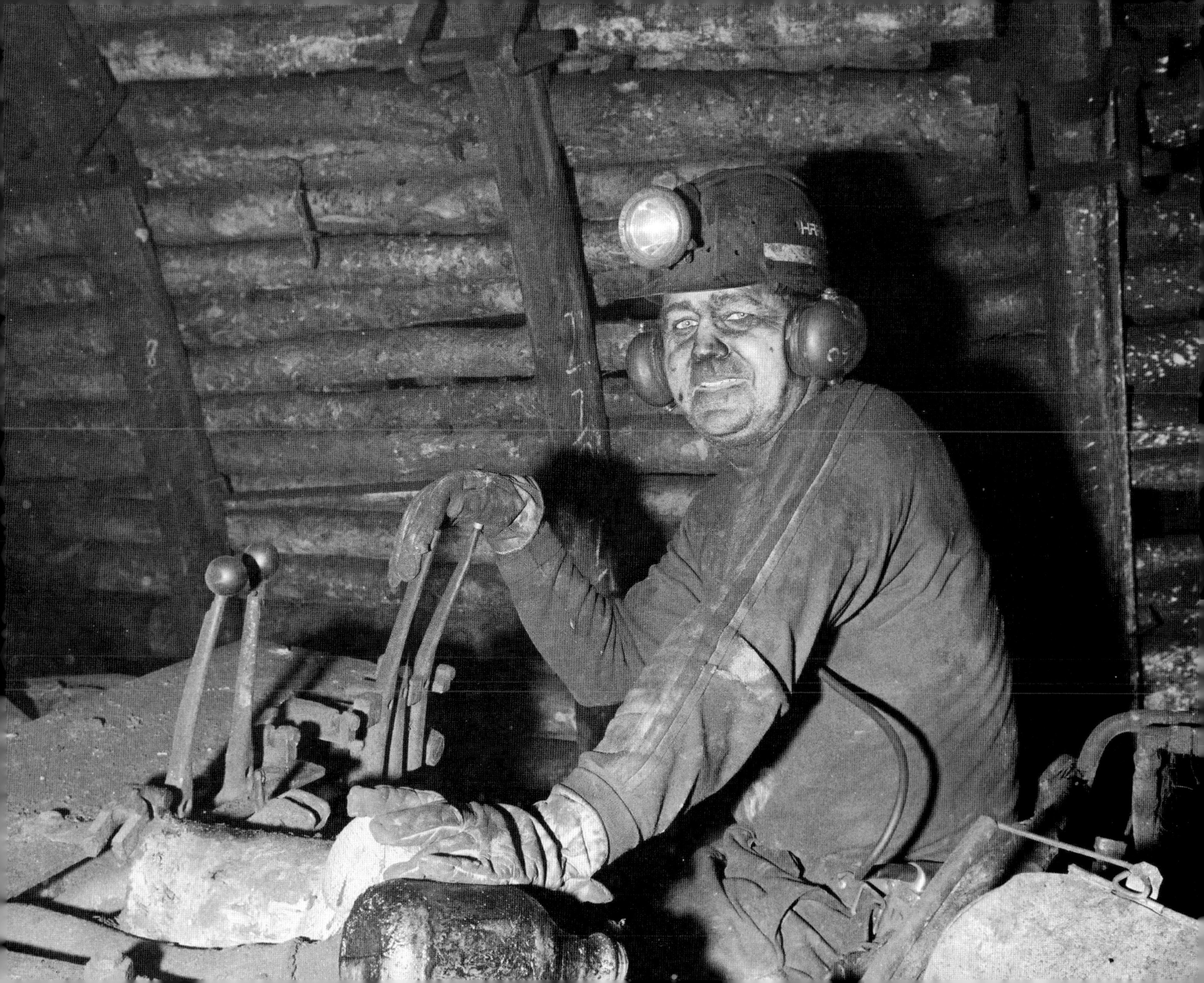

“My sister wanted me to quite the pit, so I came to the surface for awhile and I broke horses for Dominion Coal Company. We used to bring them in from the west, about 30 in a car … and that was my job. If they held a high head, they were no good for the pit.... Then I worked the stables for a while on the surface. After that I wend back in the pit.

Tape-2163 25 July 1984.

In 1977 I was underground in No. 26 colliery to take photographs of “Laura,” the last pit pony to work in the Cape Breton coal mines and perhaps the last pit pony in North America. Laura began working in the colliery in 1972 and was stabled at the pit bottom, some 900 feet underground and was cared for by Kenny Seymour.

Cape Breton coal mines used pit ponies for more than 250 years and they were most common in the 1930s and 1940s. By 1947, there were 100 pit ponies in the No. 26 colliery. Due to the mechanization of the mines, the use of pit ponies diminished over time.

Owen

Last Pit Pony – Laura, 1975

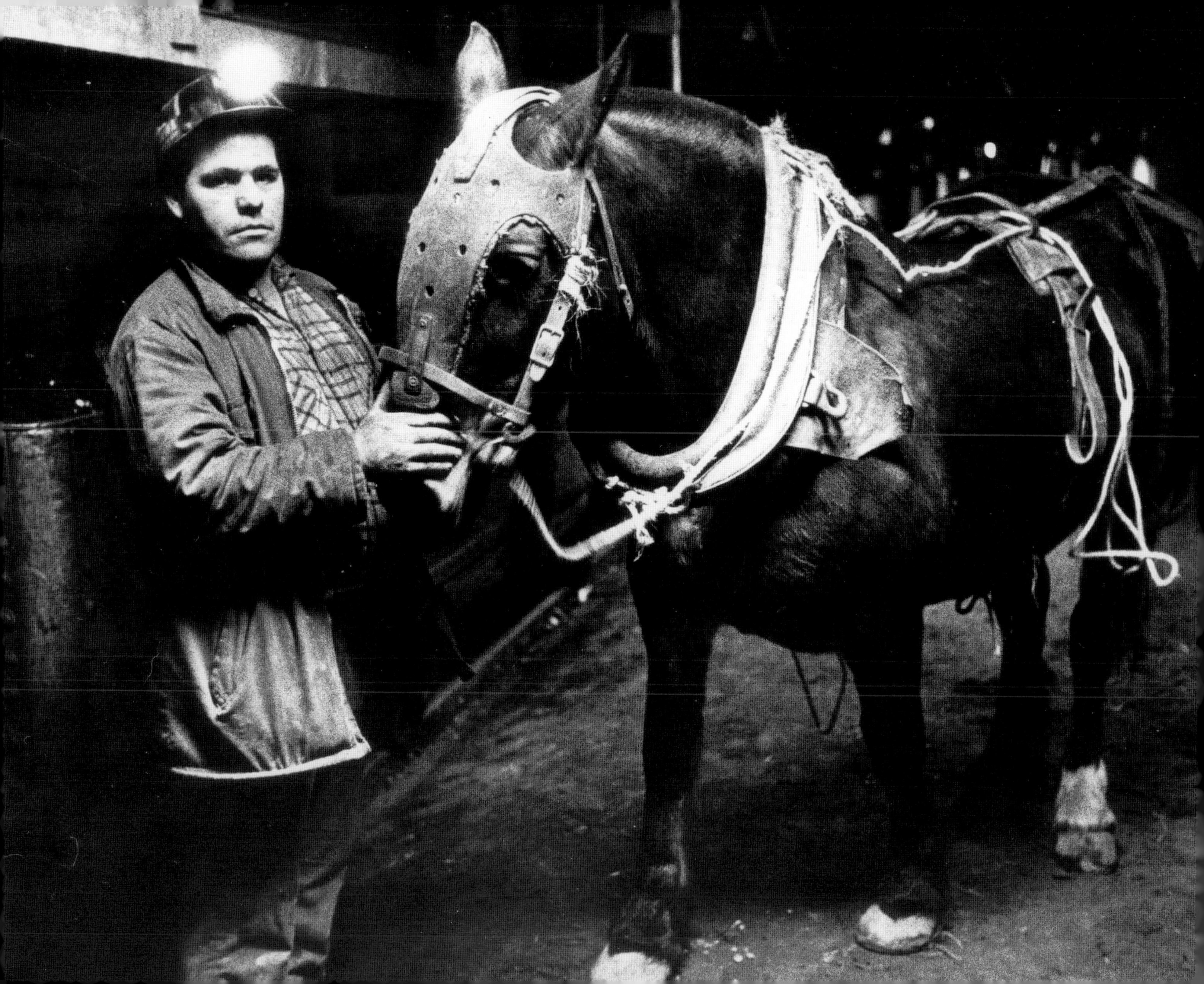

144-A-9 No. 26 Colliery 1981

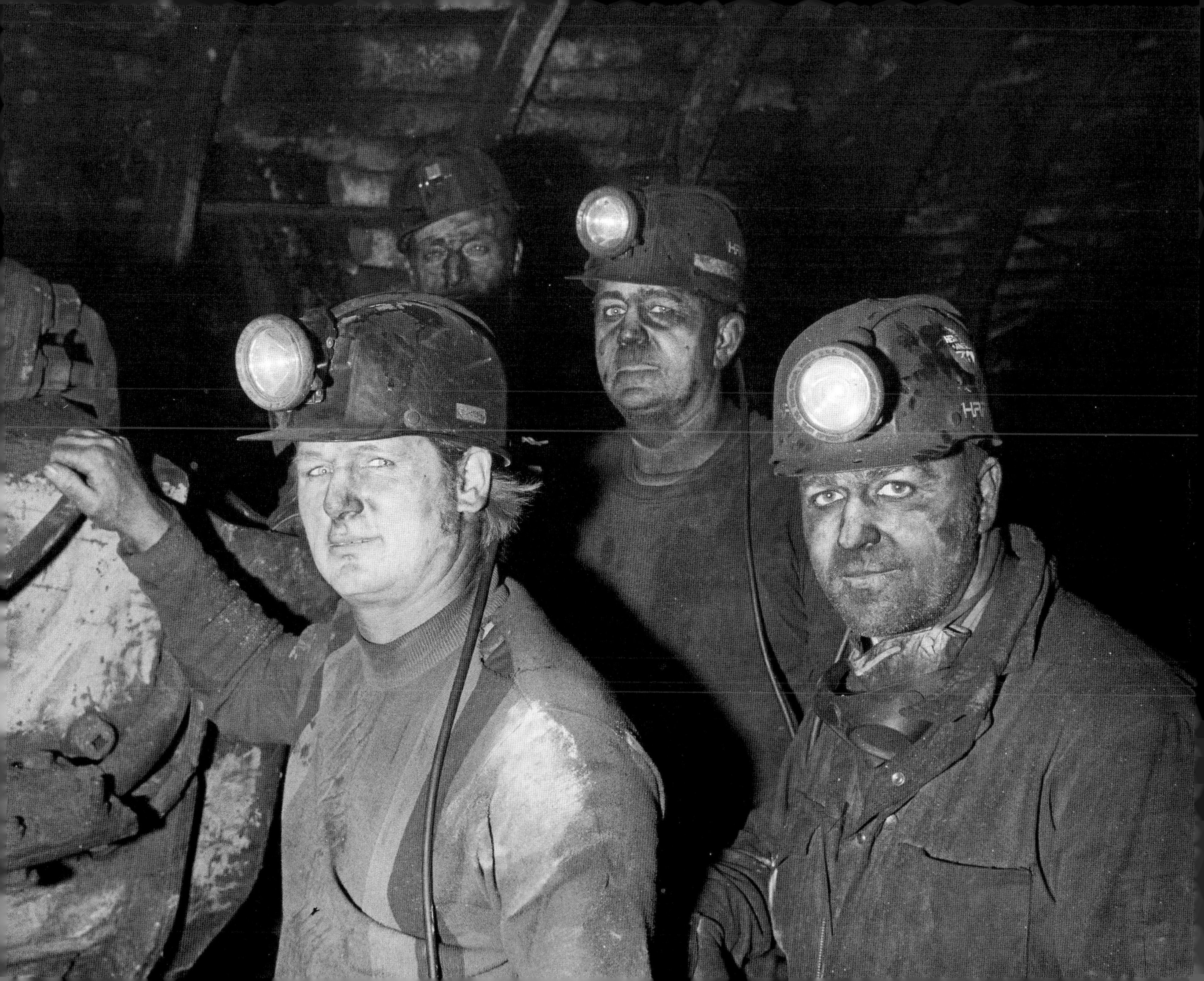

That's a gas pump, used to pump out the methane gas.

144-A-7 No. 26 Colliery 1981

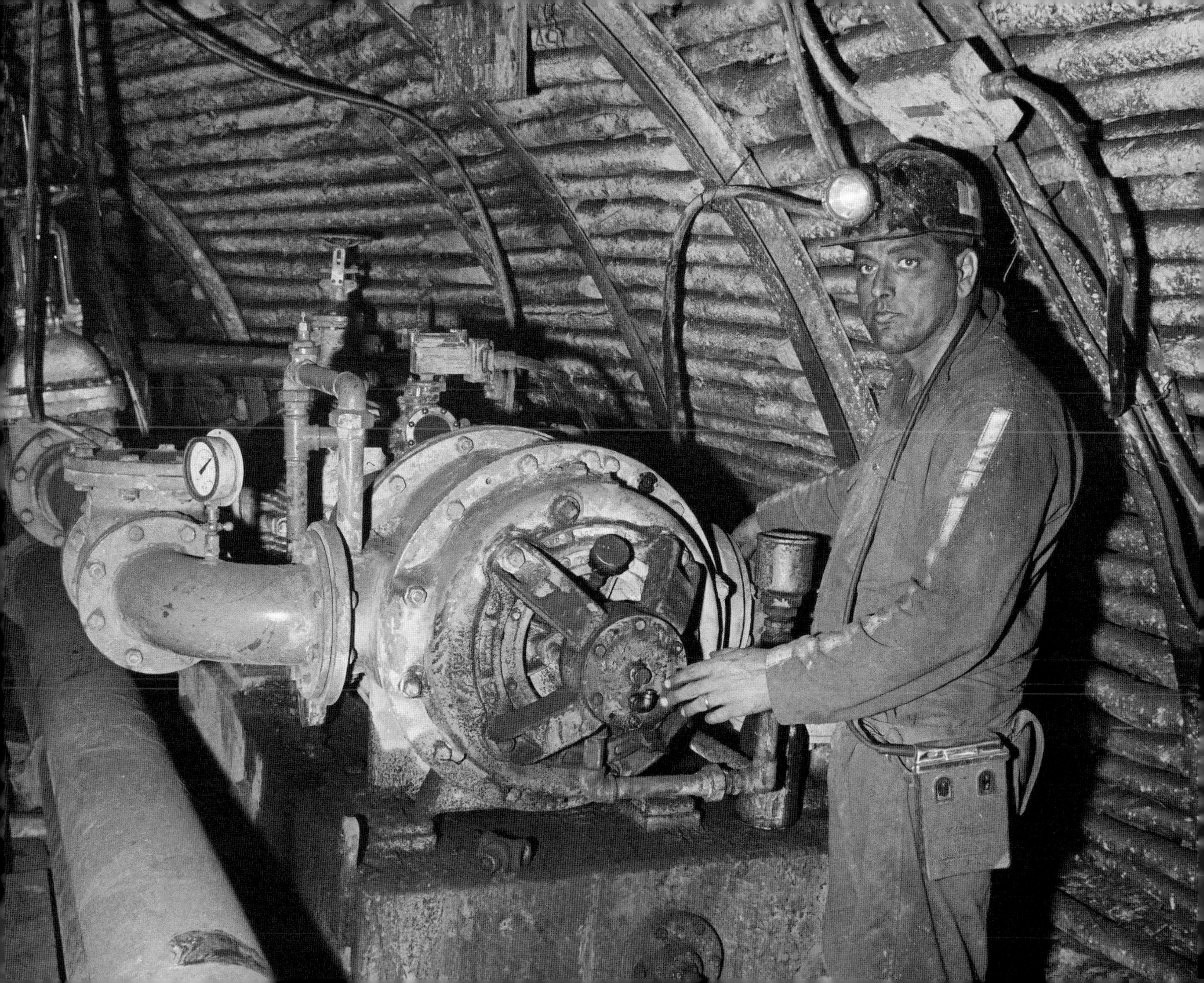

Mining can get into one's blood. Some men continue to mine elsewhere—out West, Northern Ontario.

1000-A-26 No. 26 Colliery 1985

Inset 1000-D-26 No. 26 Colliery 1985

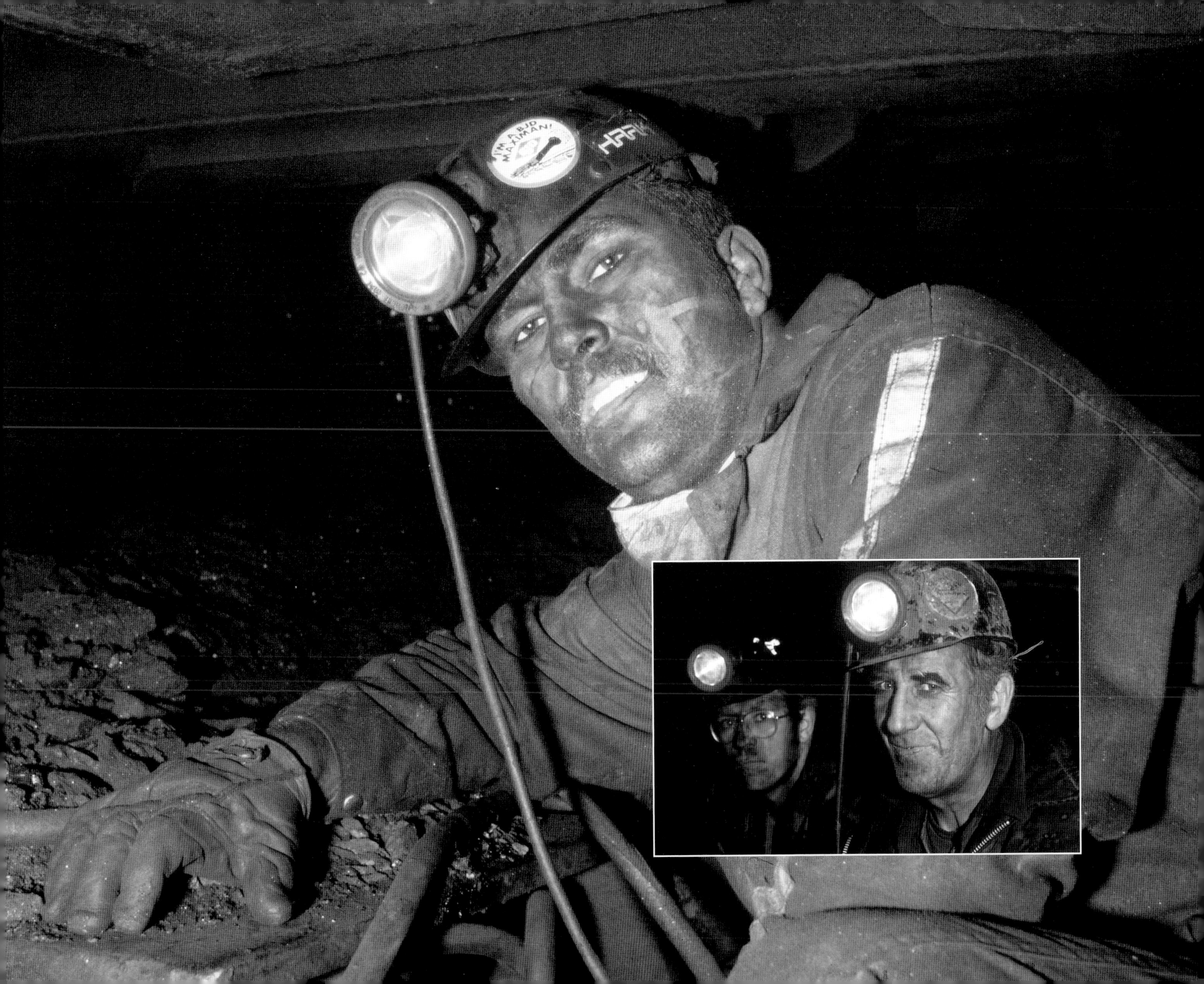
I'M A BJD MAXIMAN!

Wearing stone dust.

1000-D-3 No. 26 Colliery 1983

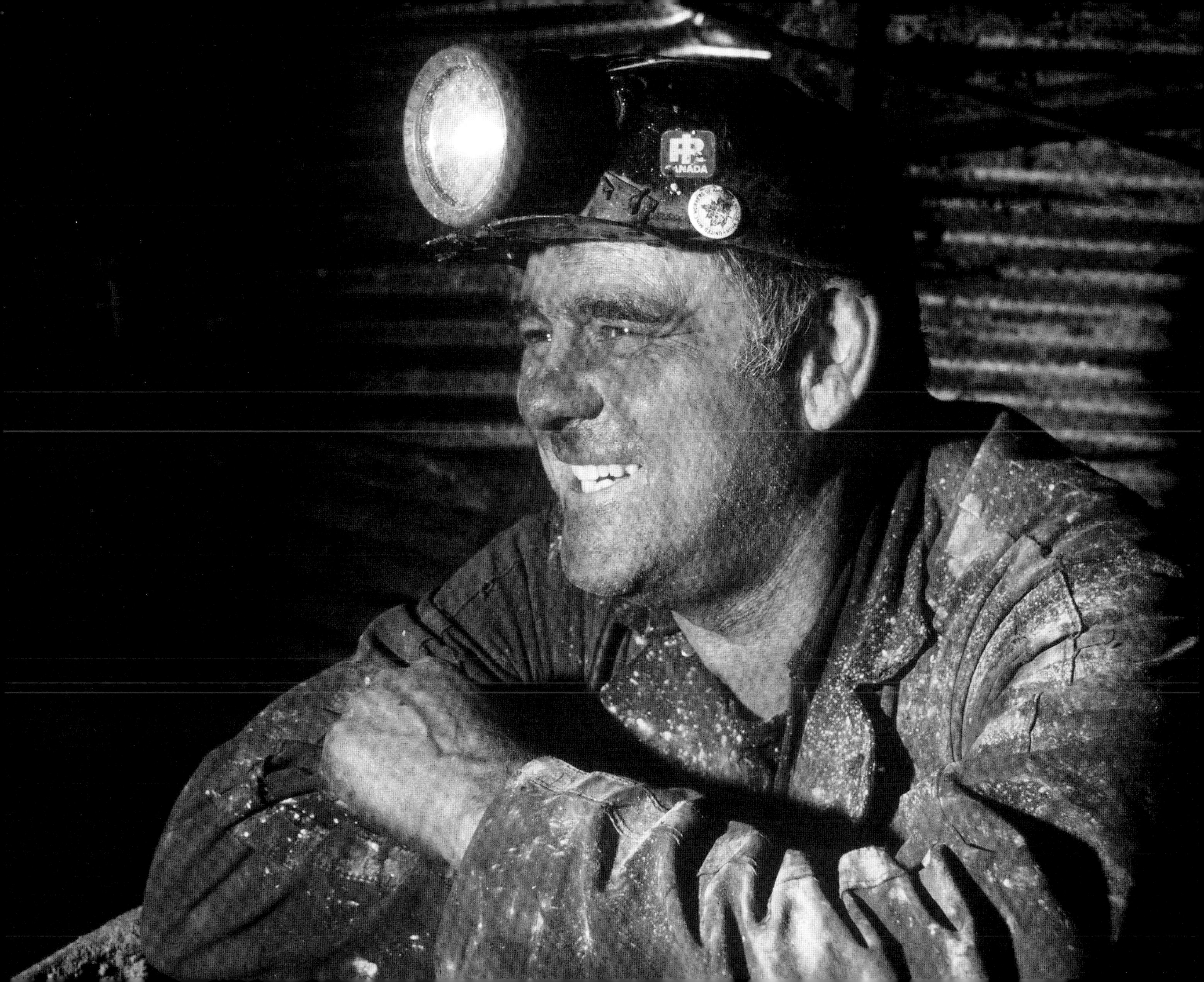
CANADA

The aluminum lunch cans were eventually banned from the mine, due to potential for sparking. Most of these cans were made by a hoist operator, who made them while on shift.

The Big Cage, meant to hold 30, sometimes as many as 80-90 would get in.

Everyone was racing to get into the shower; if you didn't you had to wait.

159-A-1 No. 26 Colliery 1977

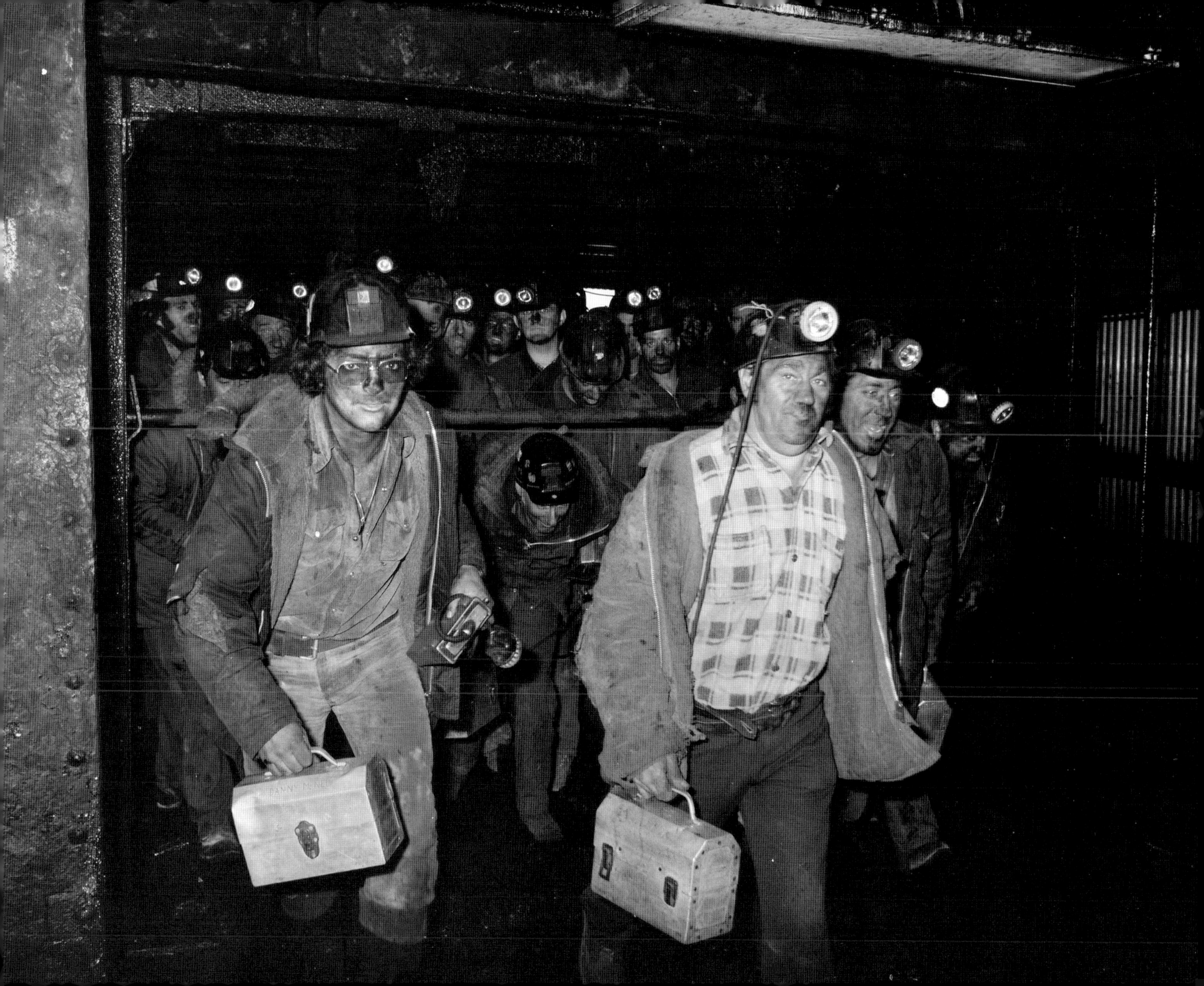

Race to the showers—race to have a cigarette.

No. 26 Colliery – miners coming out of cage, 1975

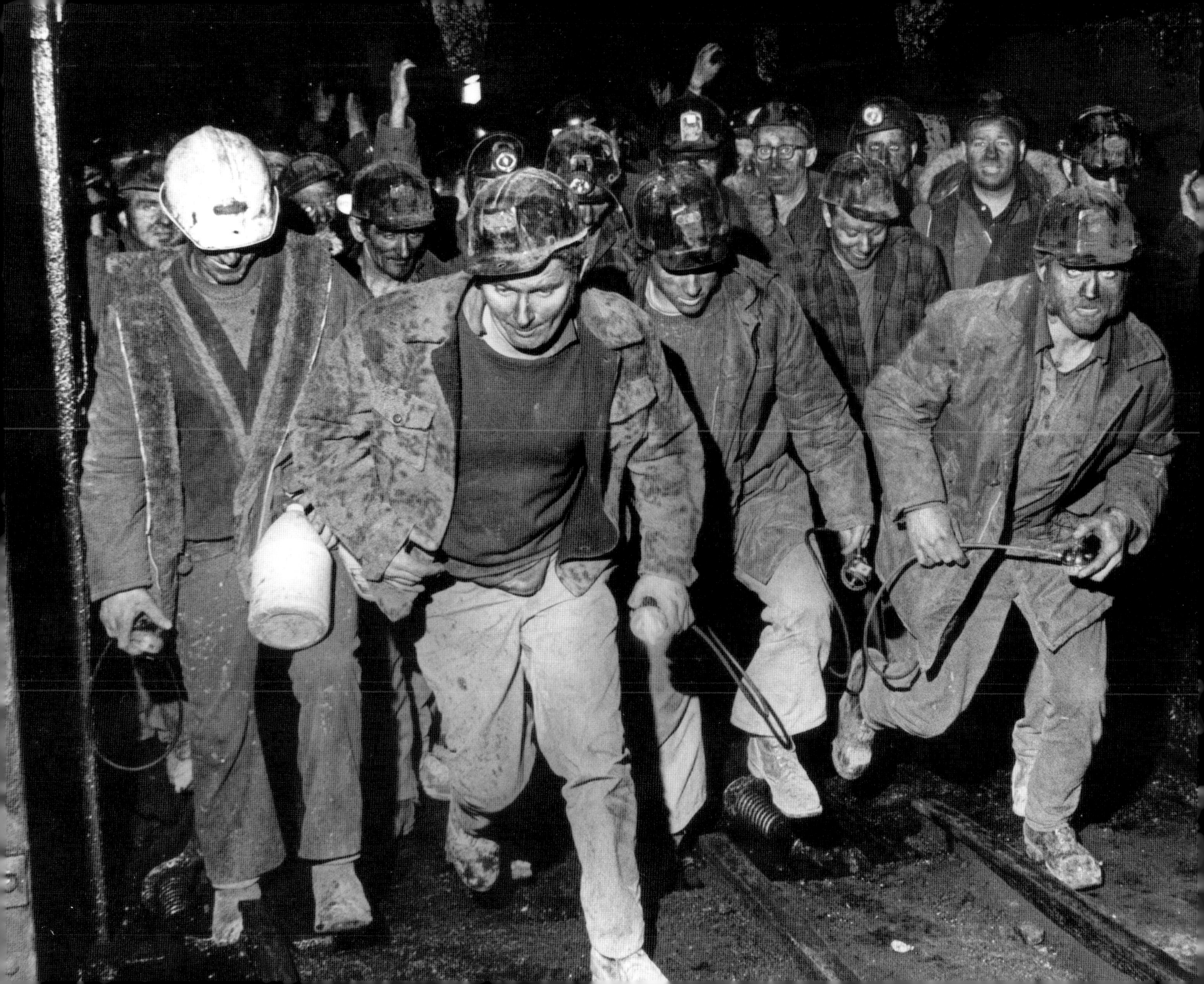

279-A-19 Lingan Mine 1974

Lingan Mine

1974
1975
1981
1983
1985
1986

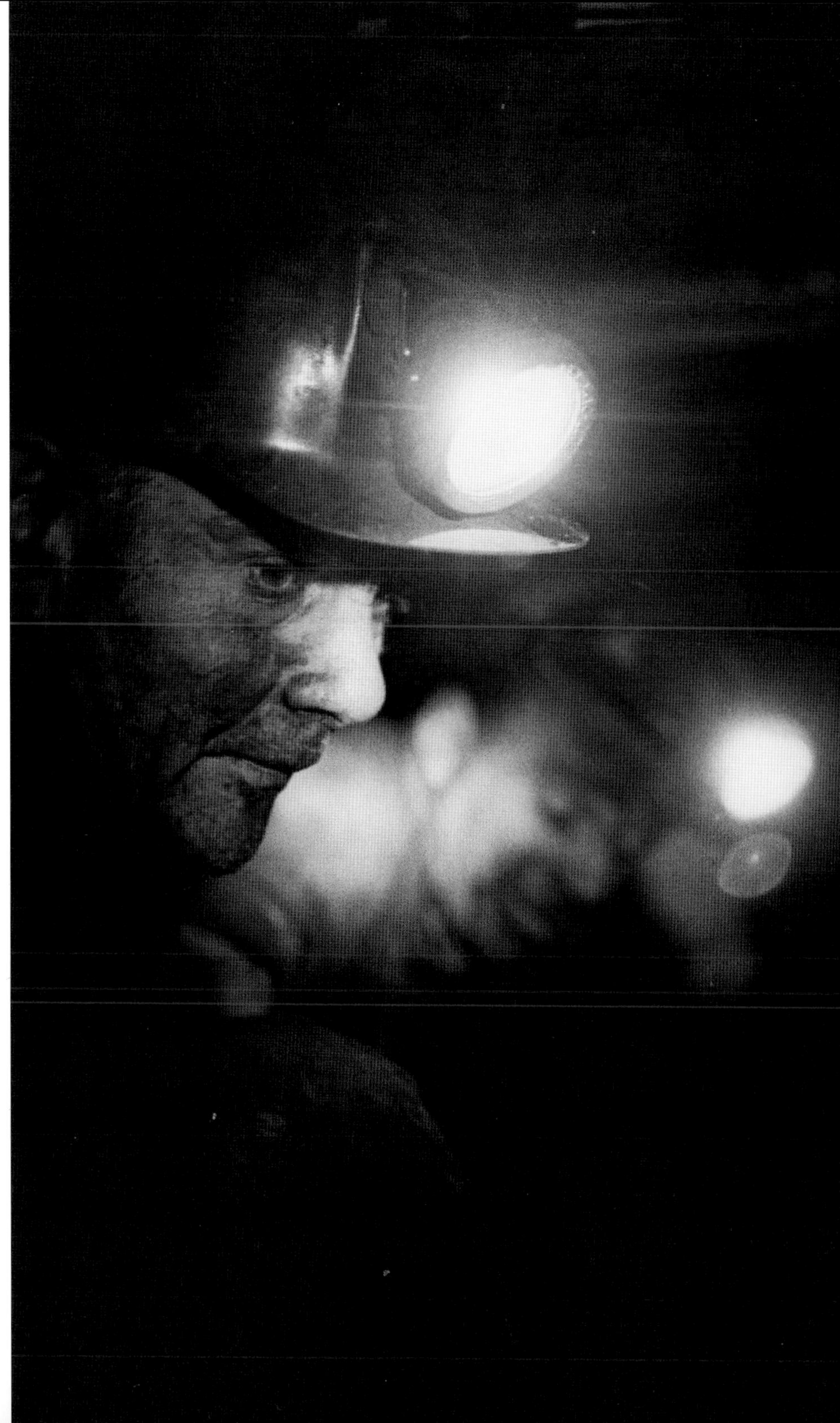

Top 280-A-29A Lingan Mine 1975

Bottom 279-B-19A Lingan Mine 1974

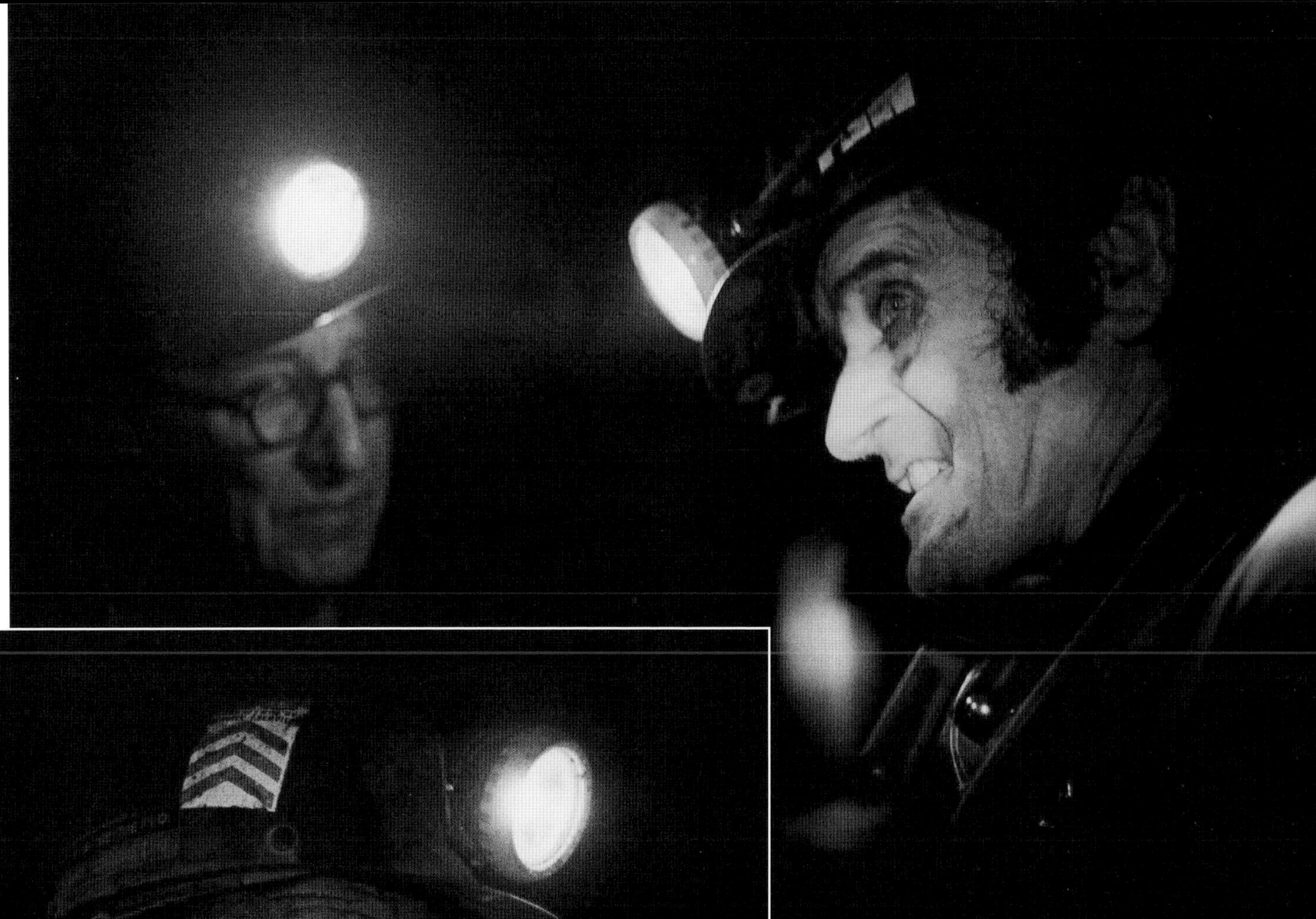

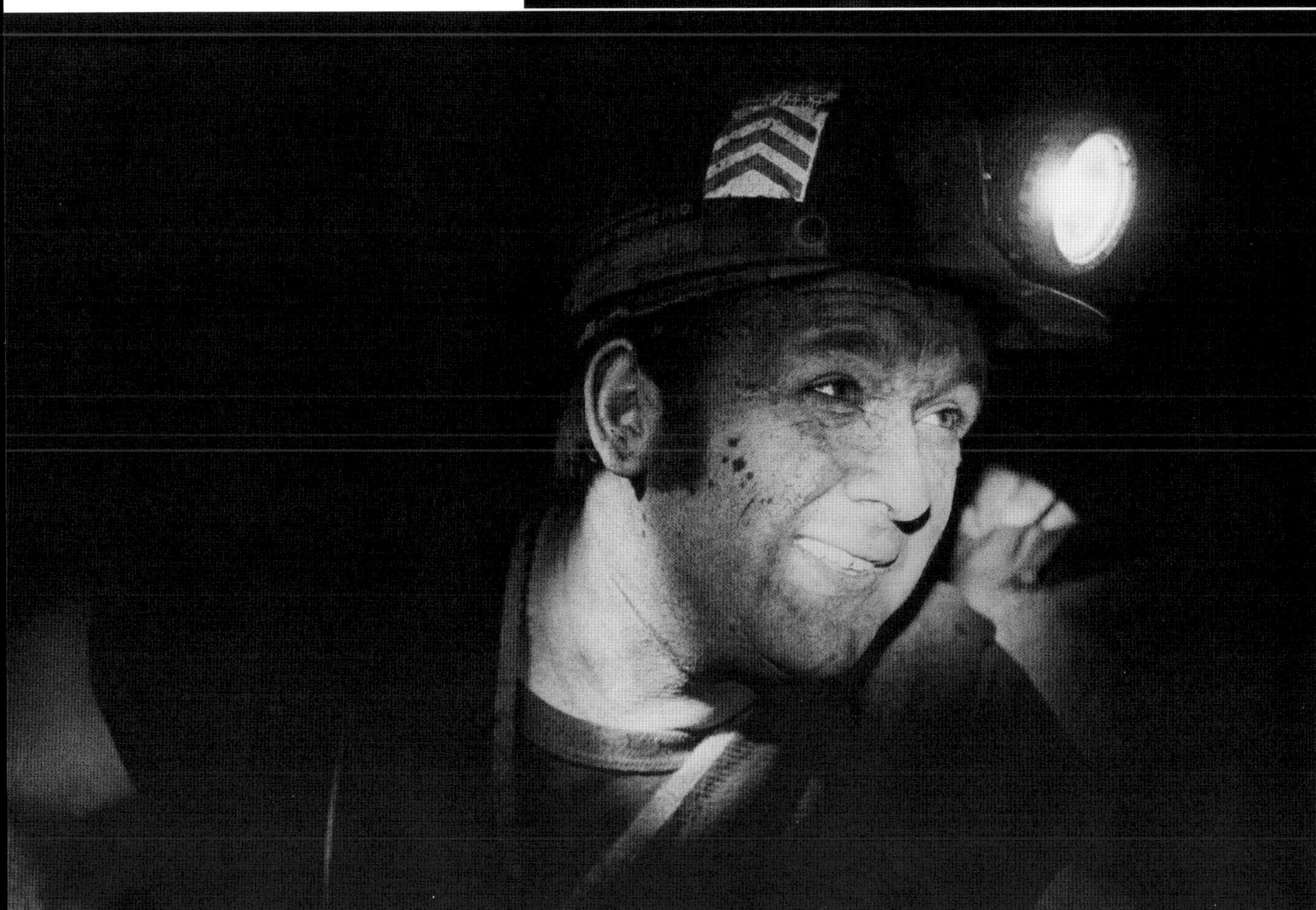

The pits today are wonderful pits, great ventilation. The pit is a good place to work today. Big difference when I was working.

Tape-2163 25 July 1984.

After the explosion in twelve, the government stepped in fixing up the mines and everything, fixing up the airways, getting ventilation. And the mines are the best in the world now, yeah, good conditions and everything.

Tape-2163 25 July 1984.

"Air bag" overhead—ventilation duct.

330-C-33 Lingan Mine 1986

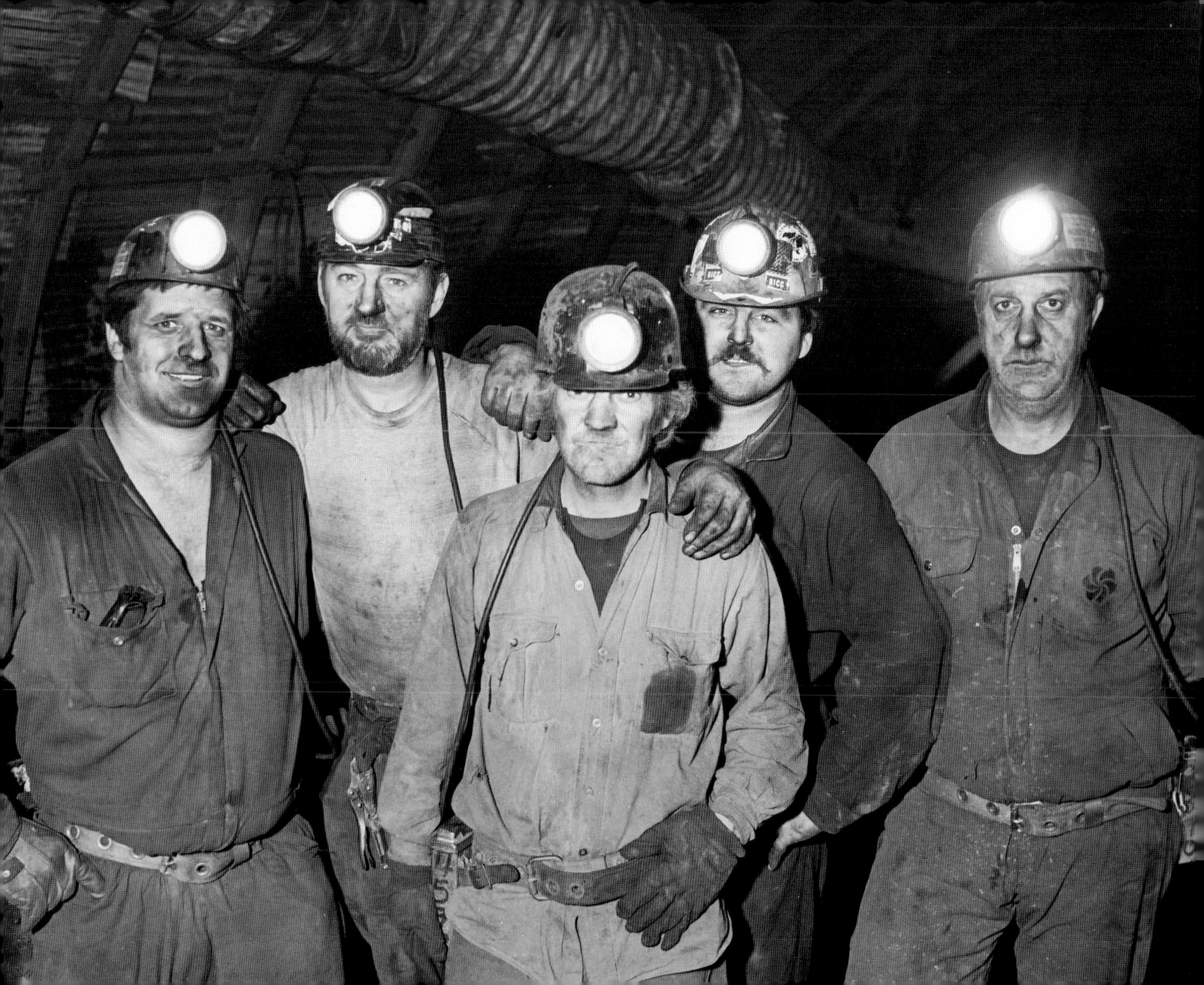

“Lamps belonged to the Company. Before that they had lanties. They were lamps with oil you had to light.

Tape-2146 20 August 1984.

The first fatalities at Princess took place on May 21, 1878, when a pocket of gas was ignited by a naked lamp. The resulting explosion killed six miners. During the early years of mining, there were a great number of gas explosions caused by the use of open candles for illumination. Later, with an efficient amount of ventilation maintained by powerful fans, not much inflammable gas was met with.

Owen

They are wearing lock flame safety lamps. No coveralls, just clothes from home. One has a testing-lamp for gas.”

317-B-30 Lingan Mine 1985

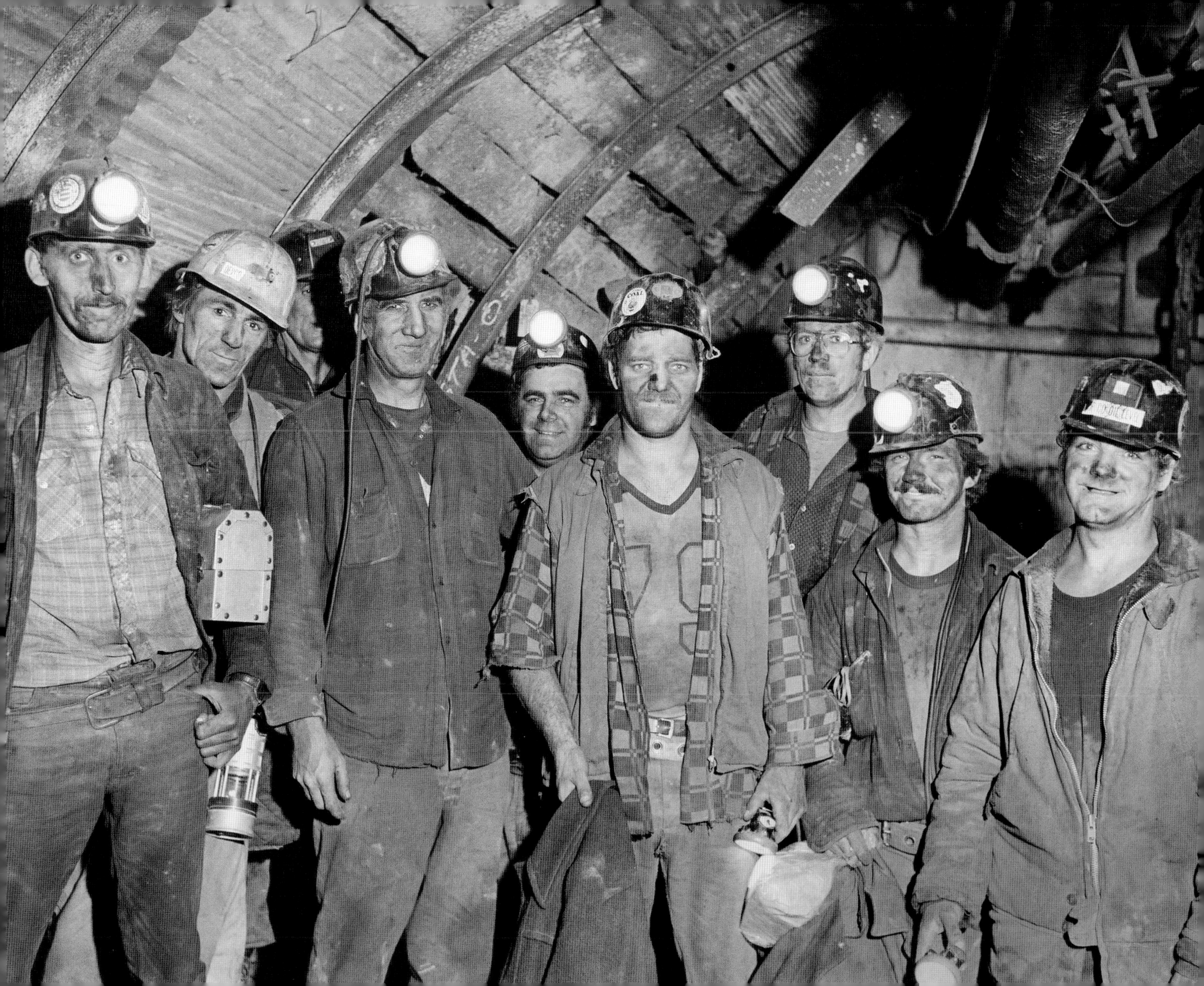

This is an Anderson 300 on the west wall, Lingan.
The coal was lower at Lingan, much less room.

137-A-5A Lingan Mine 1981

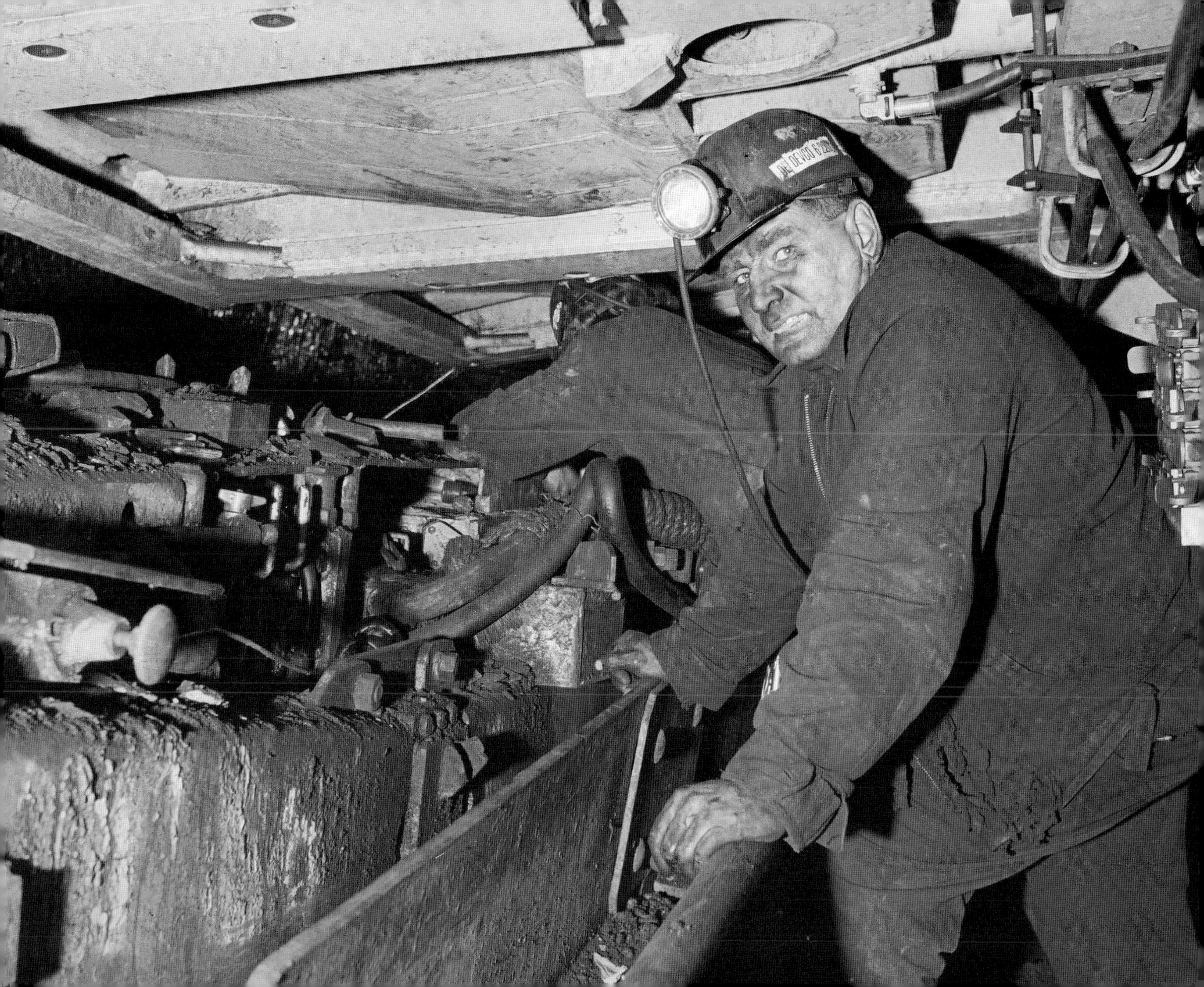
DEVICO

As the roof support drops to move forward, the "gob"—the void resulting from the removal of coal—falls behind you as seen here at the right.

These shearers operate on a haulage chain. One man was badly injured when the chain broke and hit him in the face. His nose was up on his forehead, but he was okay.

See the gouges left in the wall by the "picks" and the roof supports.

137-A-4A Lingan Mine 1981

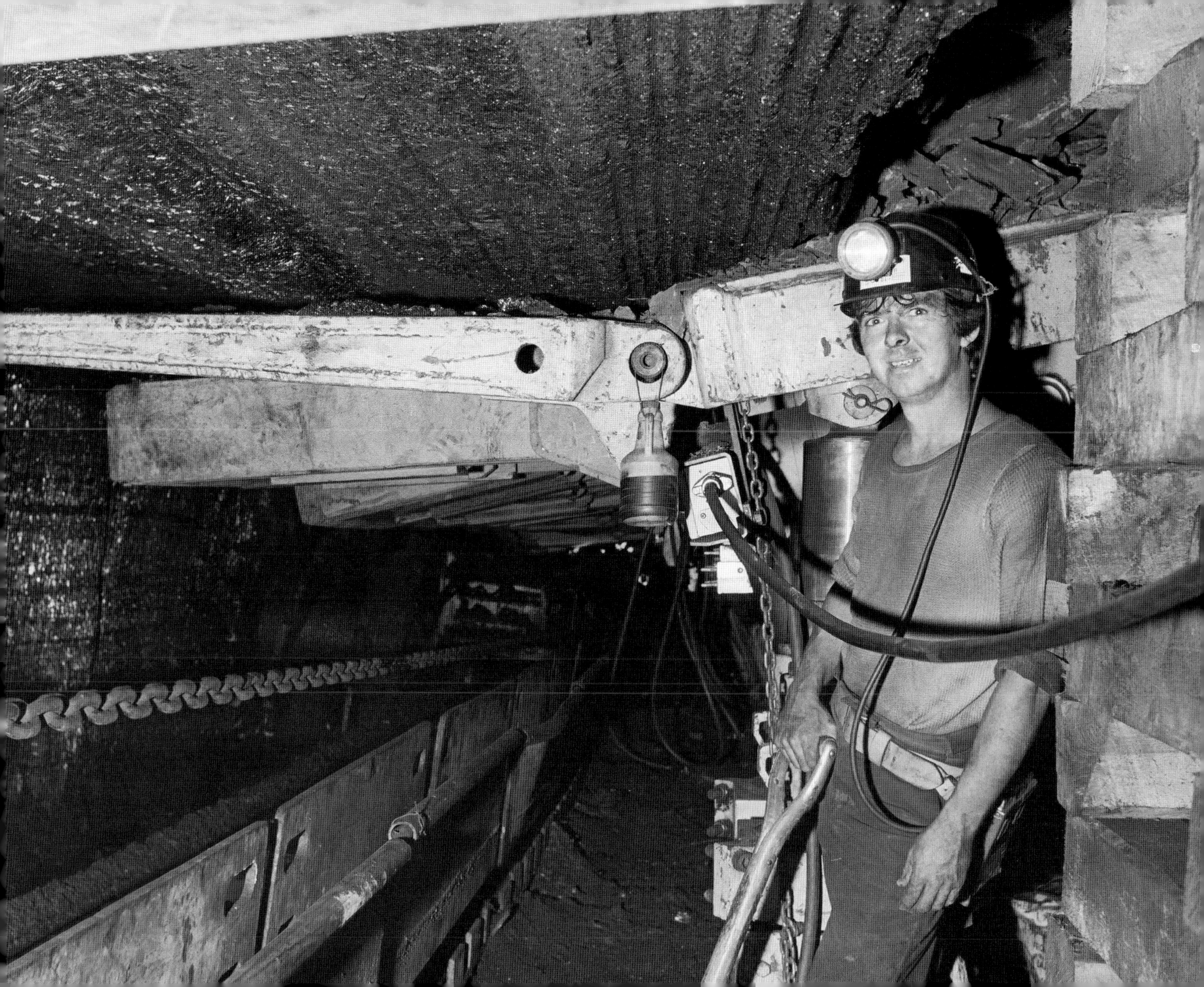

1000-C-2 Lingan Mine 1981

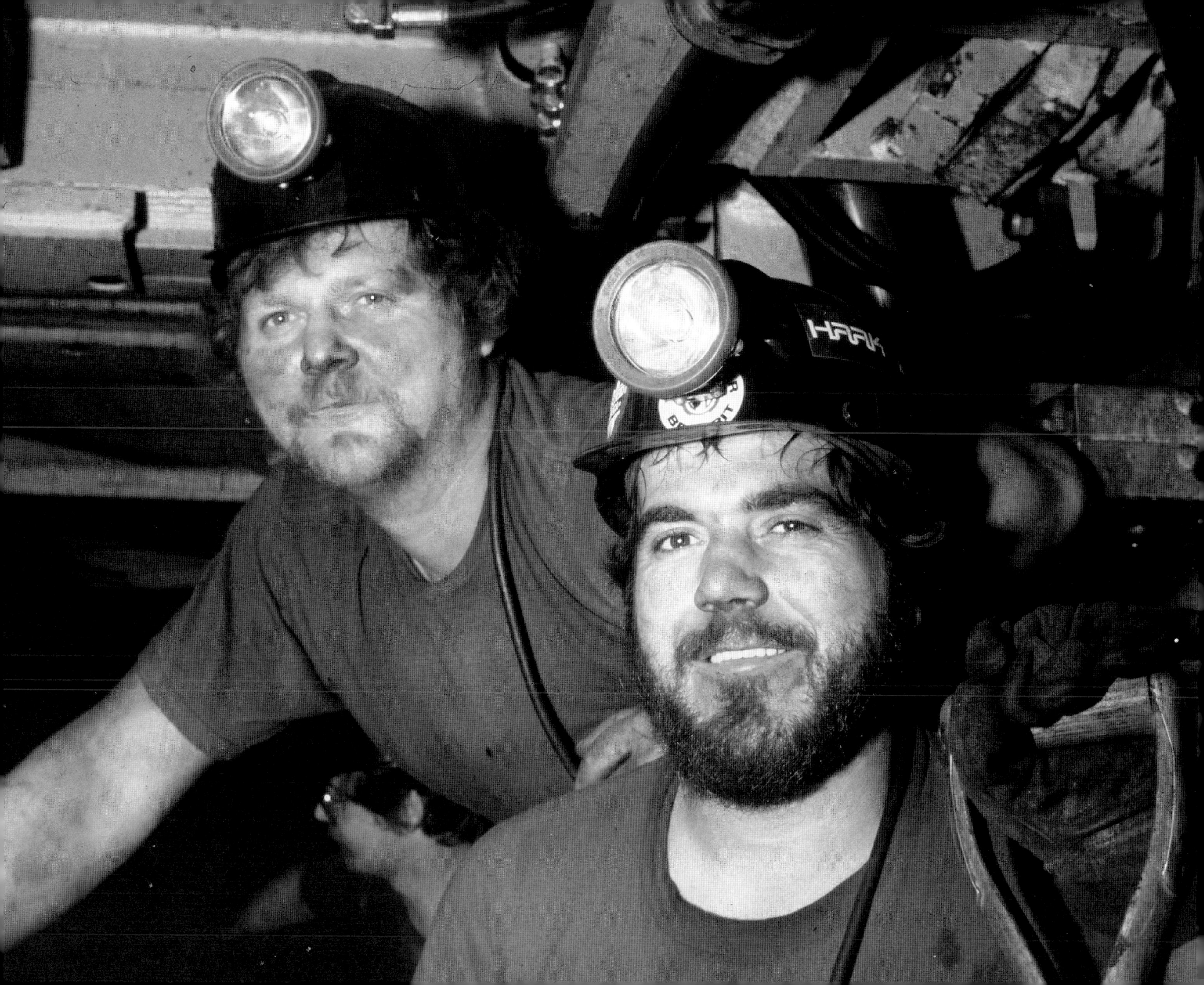

Check number on the lamp battery starts with a ‘0’, that meant they were a supervisor. Each mine started with a different number; 45 means Lingan. Seam of coal.

137-A-15A Lingan Mine 1981

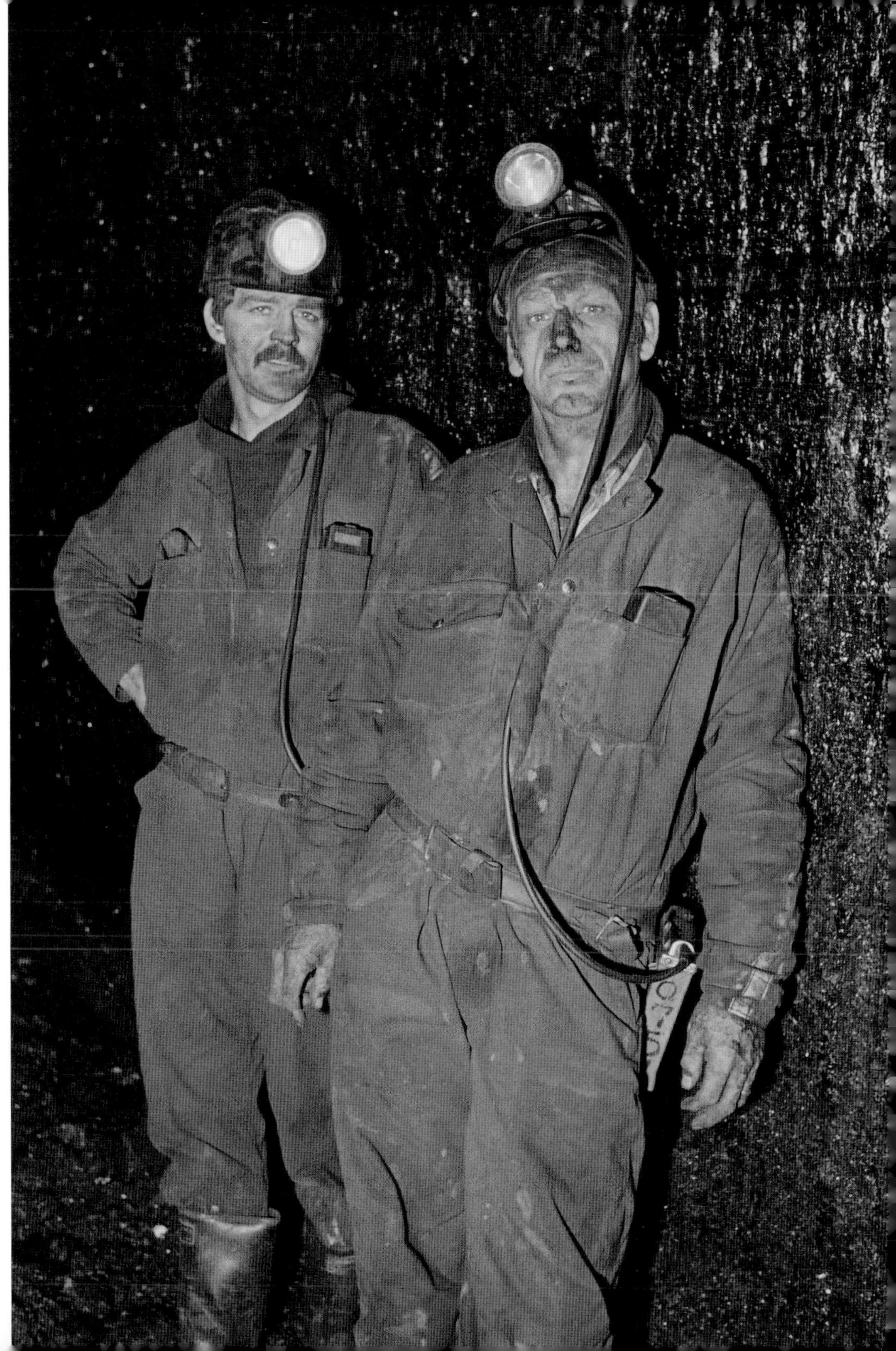

Down in the coal mines, the first concern is safety. Methane gas is explosive and common in coal mines. For this reason, the Department of Labour and the corporation, insist that any camera taken underground, did not contain a battery which could create a deadly spark. A conventional flash is not permitted either. According to the department, only a special 10 pound, steel encased, explosion proof flash, made in Germany, was permitted underground.

Owen

He's operating the roof supports. That's a methanometer in his pocket and he is wearing an older fiberglass pit hat. ”

137-A-11A Lingan Mine 1981

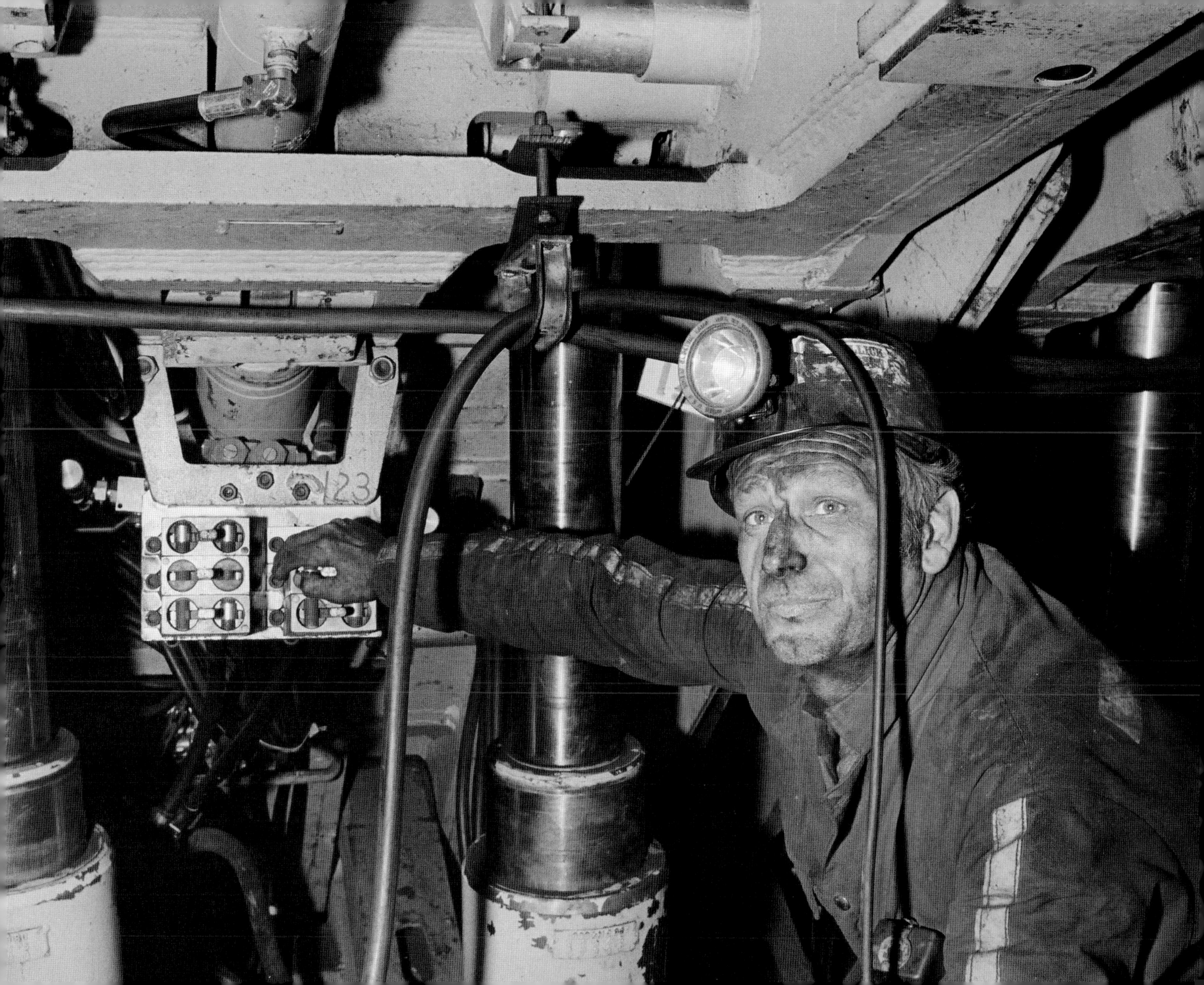

5000 psi (pounds per square inch pressure) were fed to the wall face to power the hydraulic roof supports.

They couldn't stand up all day long.

137-A-16A Lingan Mine 1981

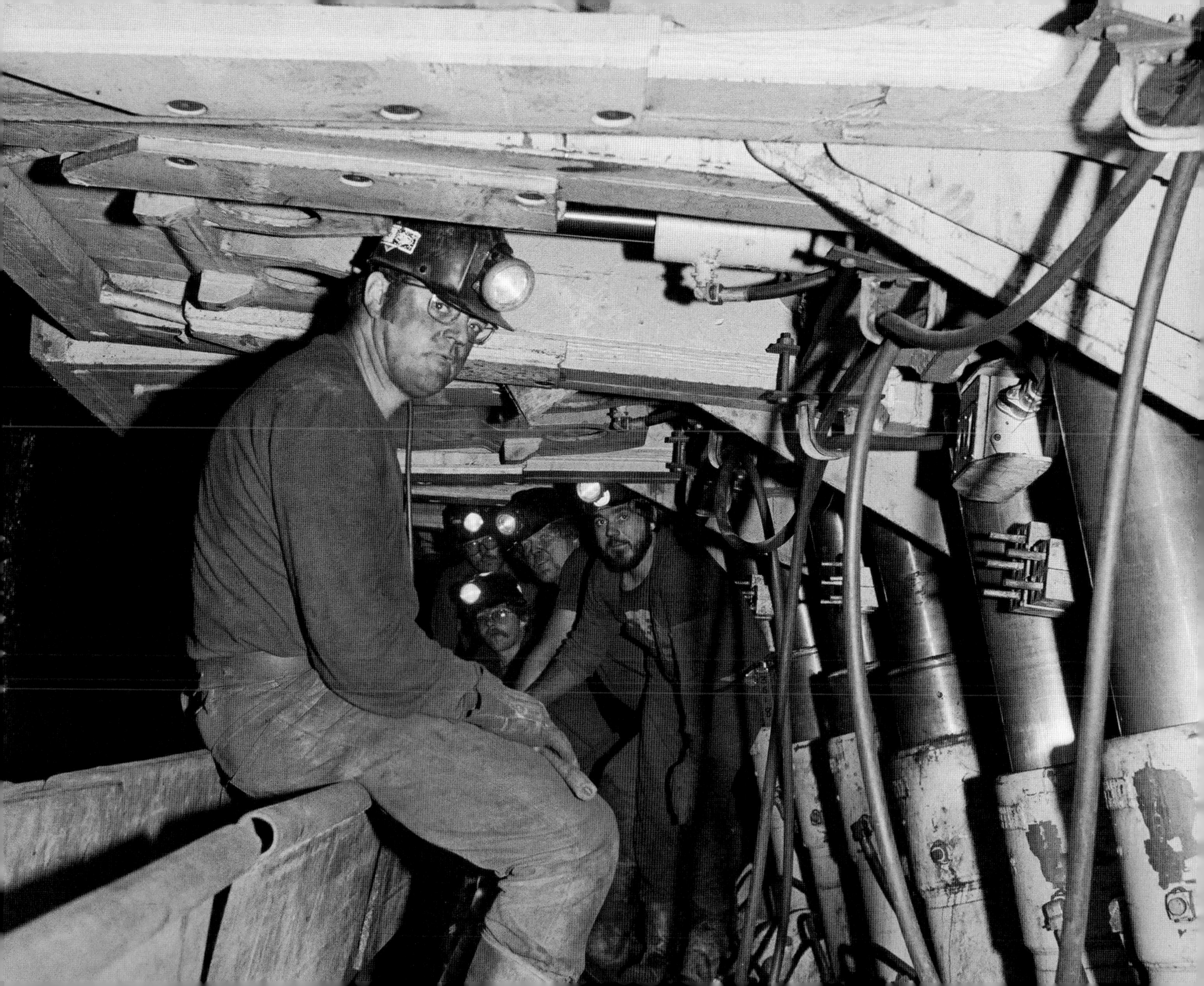

Switches were hard to manage with gloves on, so they had to bring their own "dogbone," a lever extension.

206-A-31 Lingan Mine 1983

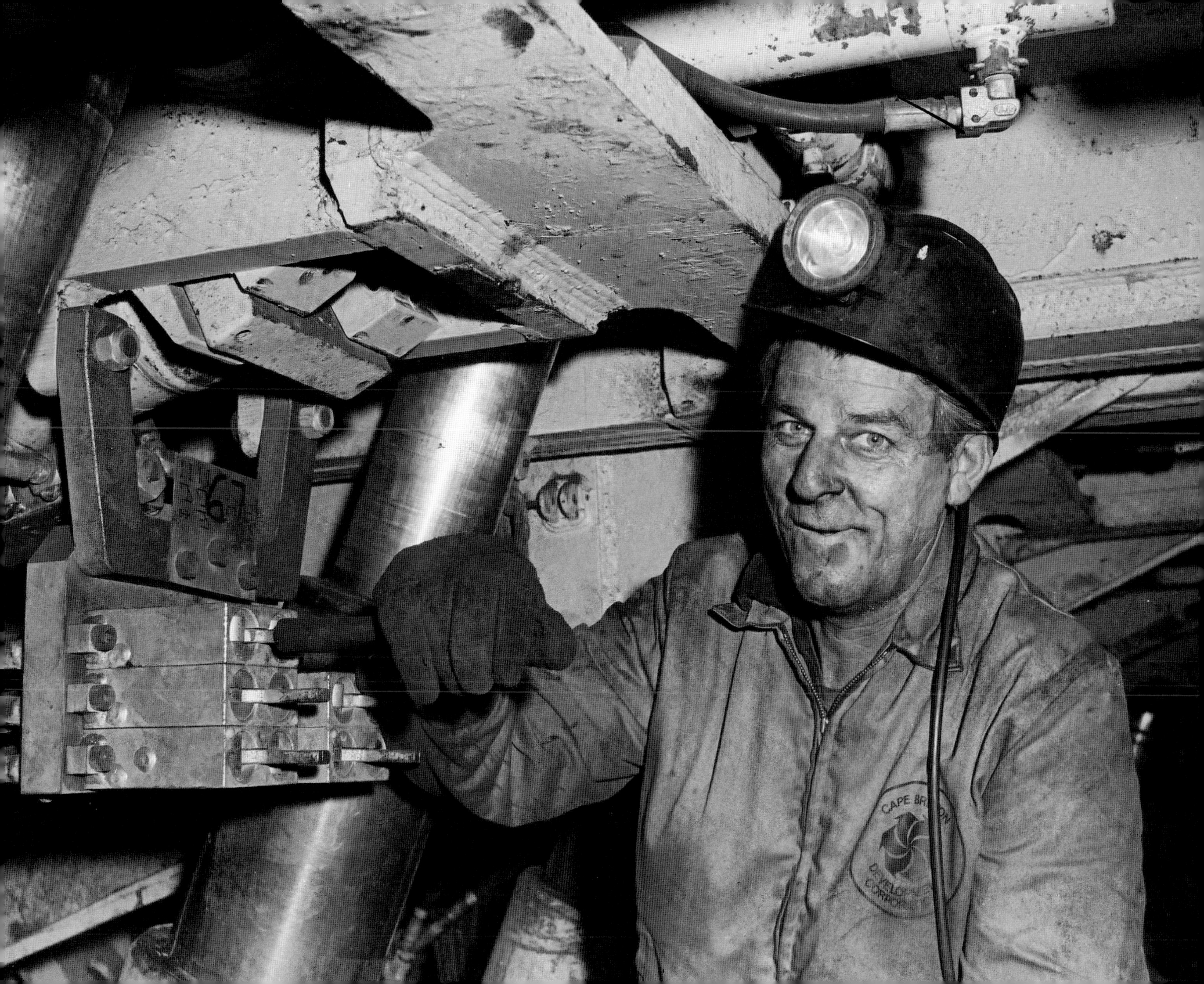

Driving the levels on the deep with a roadheader.

206-B-7 Lingan Mine 1983

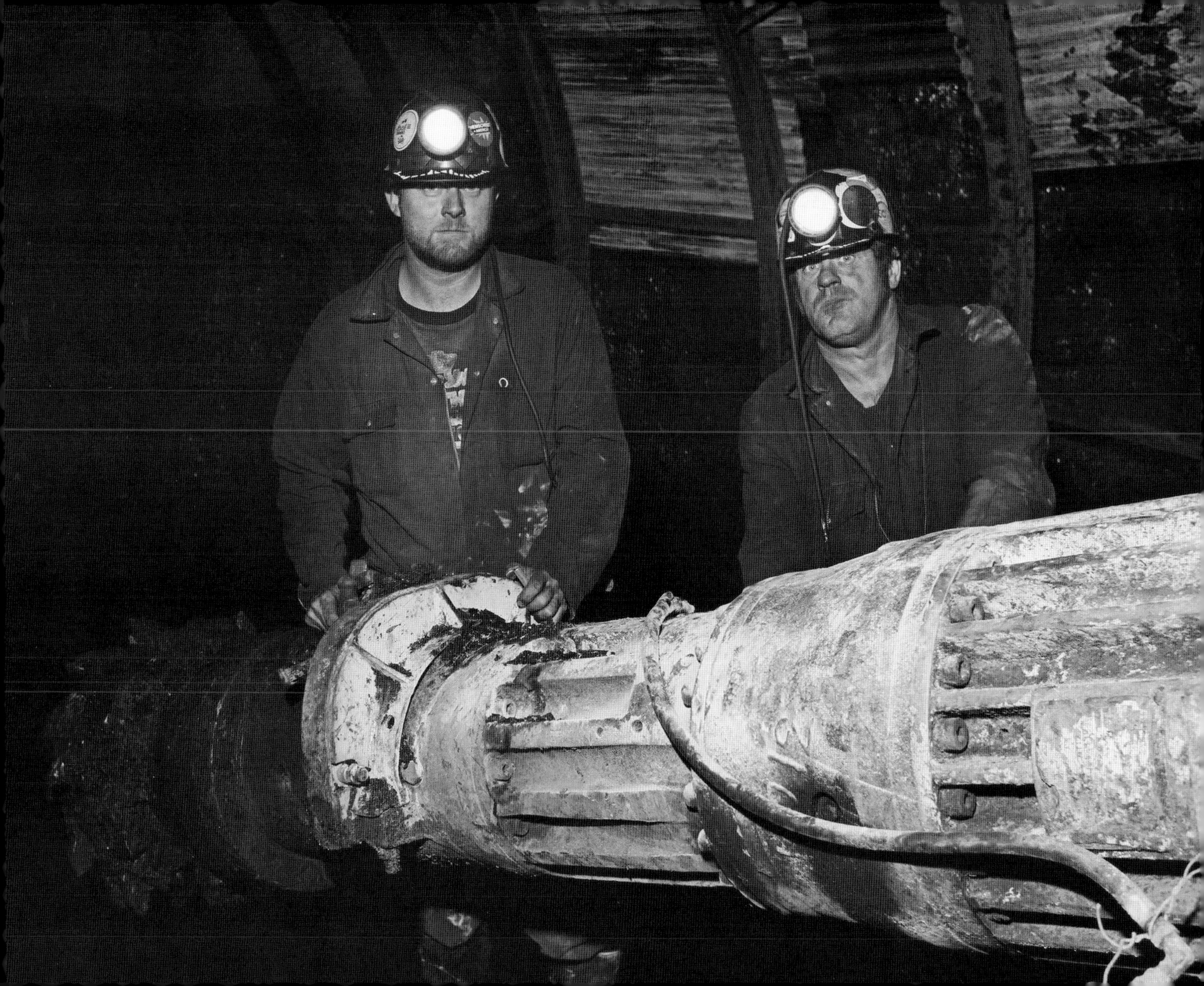

Setting up an advance wall, you drive in 200 ft. (approx. 60 m),
then develop the face, hand loading the stone.

206-A-32 Lingan Mine 1983

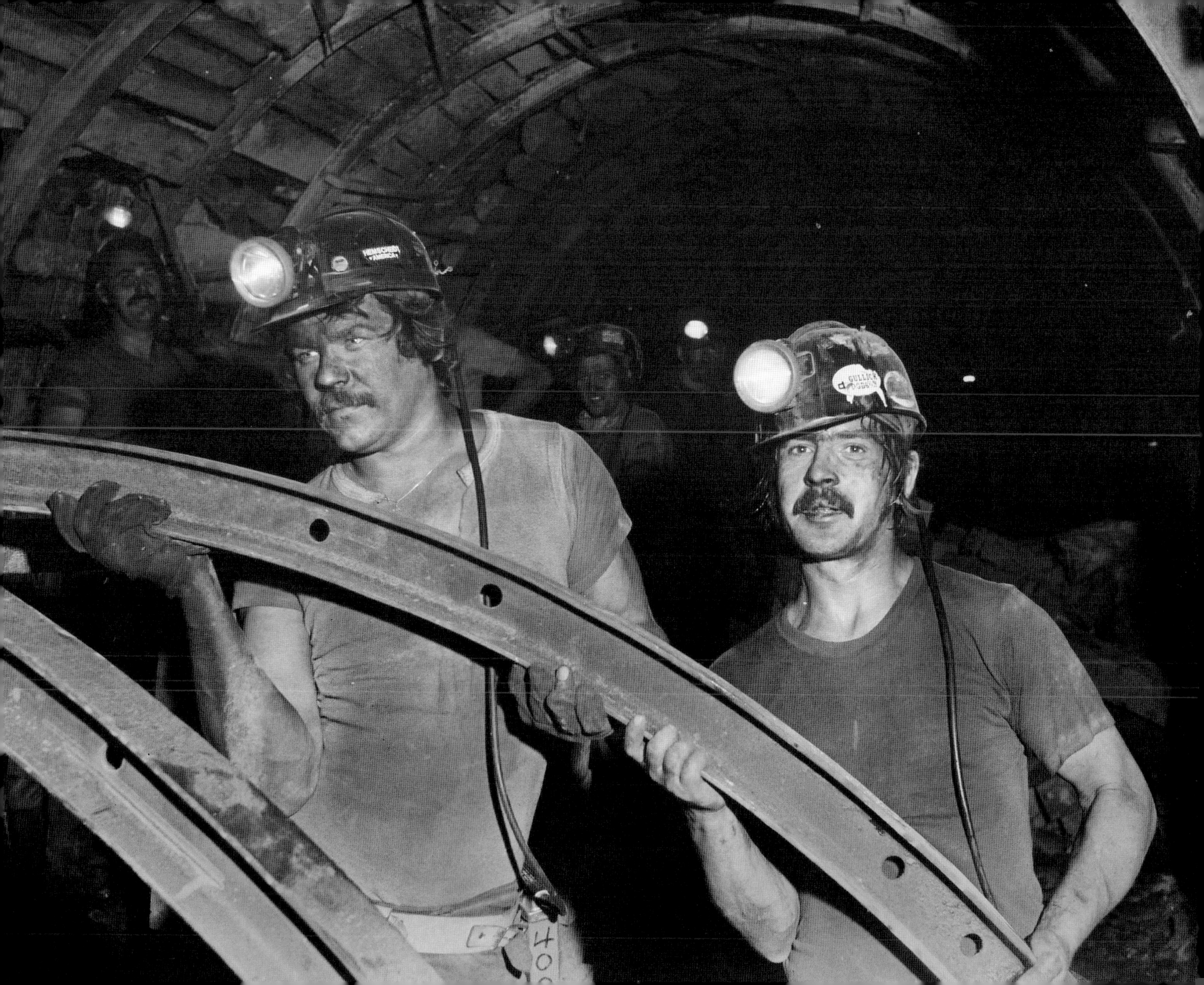
GULLICK

Console operator with the speaker box—controlled all power and communication for that wall face.

206-B-5 Lingan Mine 1983

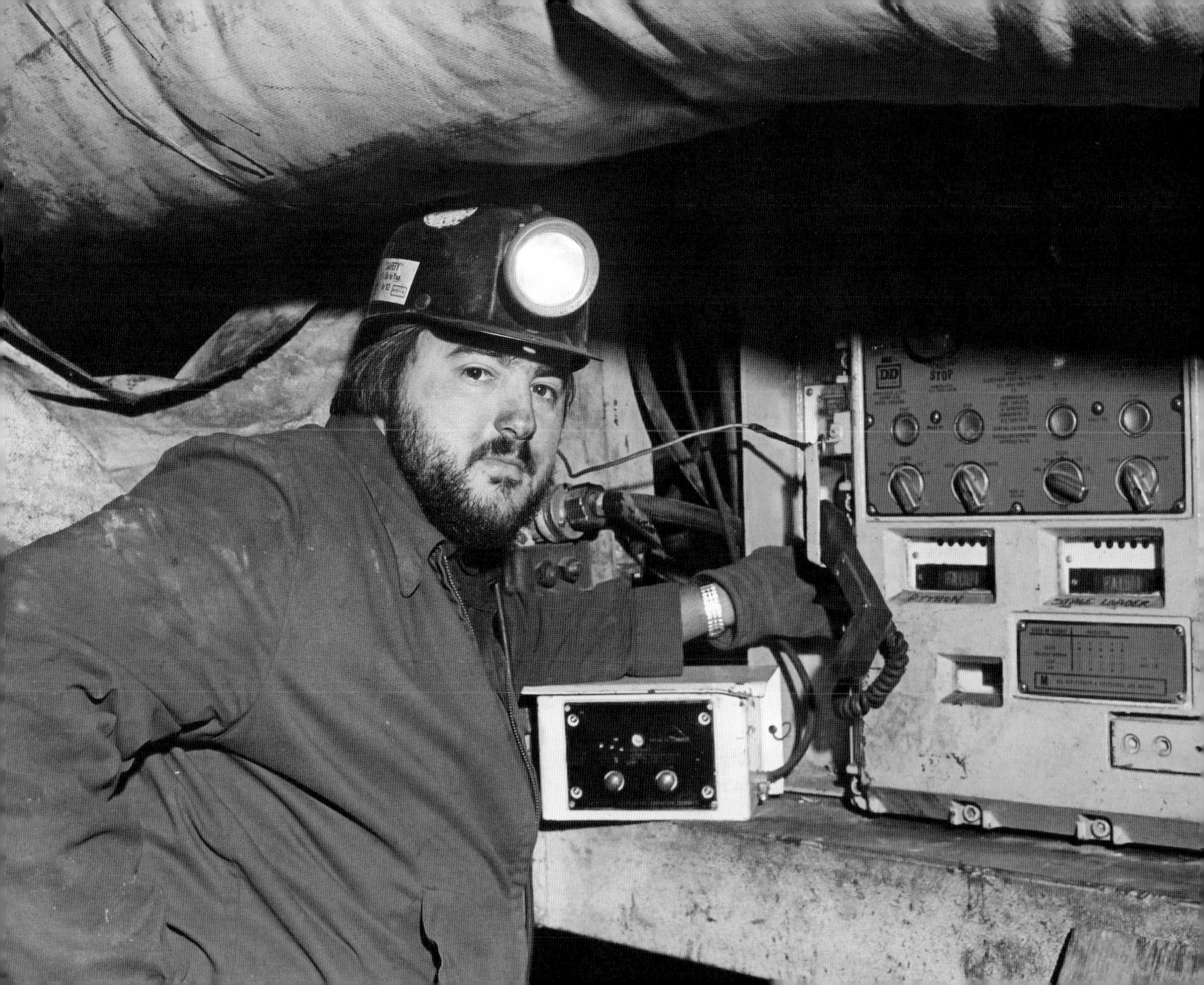
STOP

206-C-7 Lingan Mine 1983

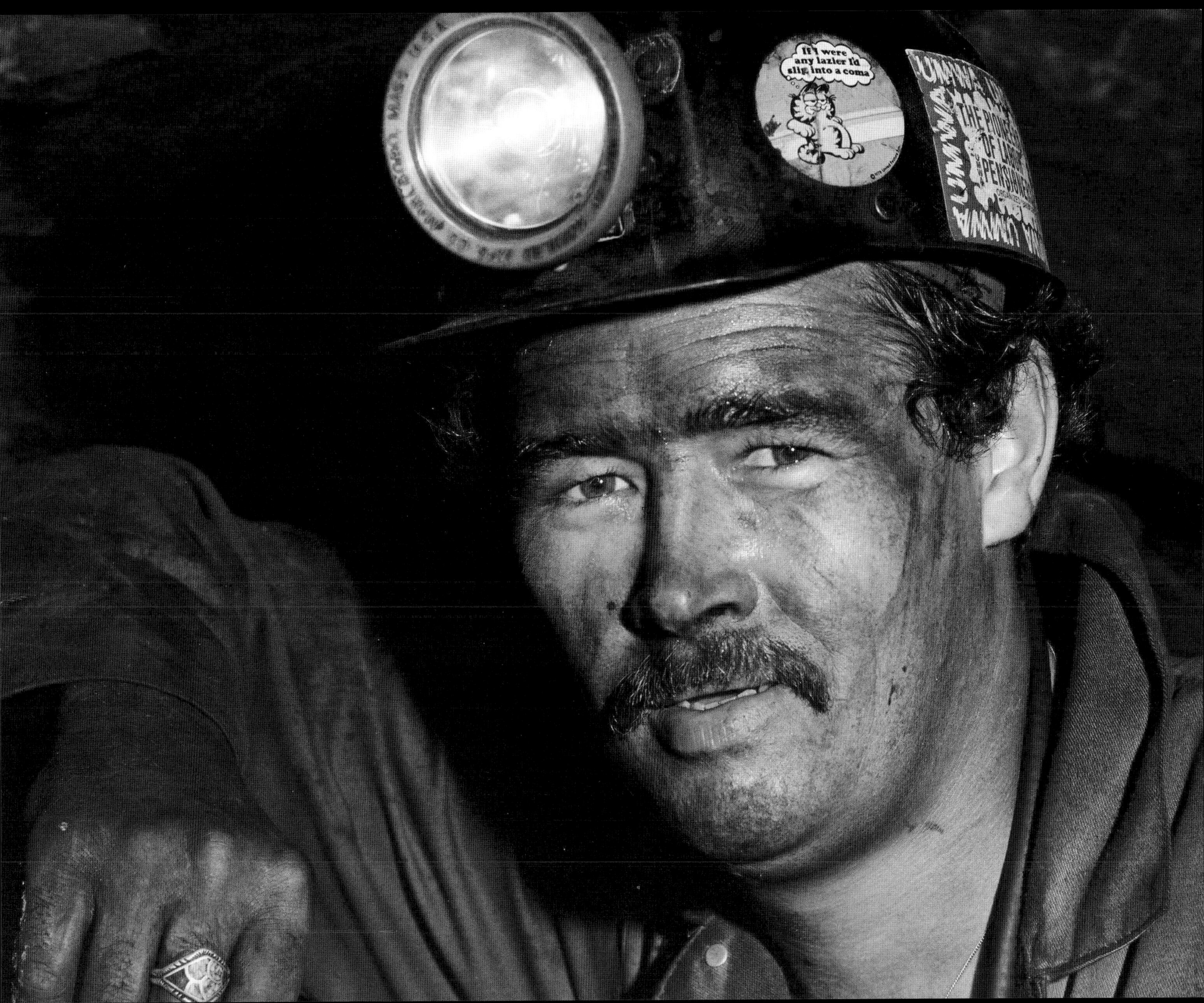

If I were
any lazier I'd
into a coma
UMWA

As far as the shovel and the pick and an axe and a saw, well that's all we ever used for timber.

Tape-2134 23 July 1984.

At the shearer. Overhead, rough cut lumber used to hold the roof.

206-A-34 Lingan Mine 1983

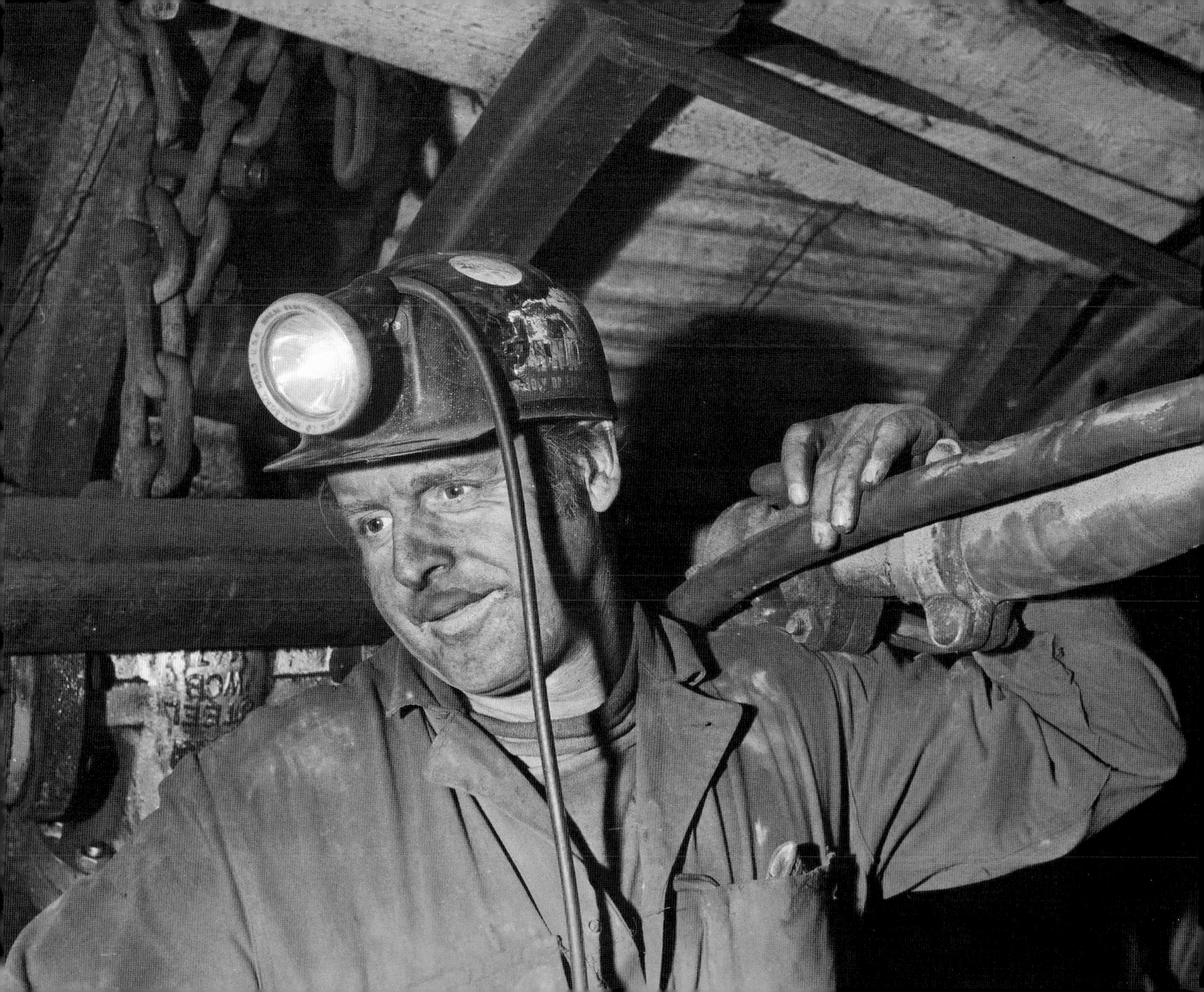

Talking on the "box," the intercom. We used to say "You're wanted on the box."

1000-D-4 Lingan Mine 1983

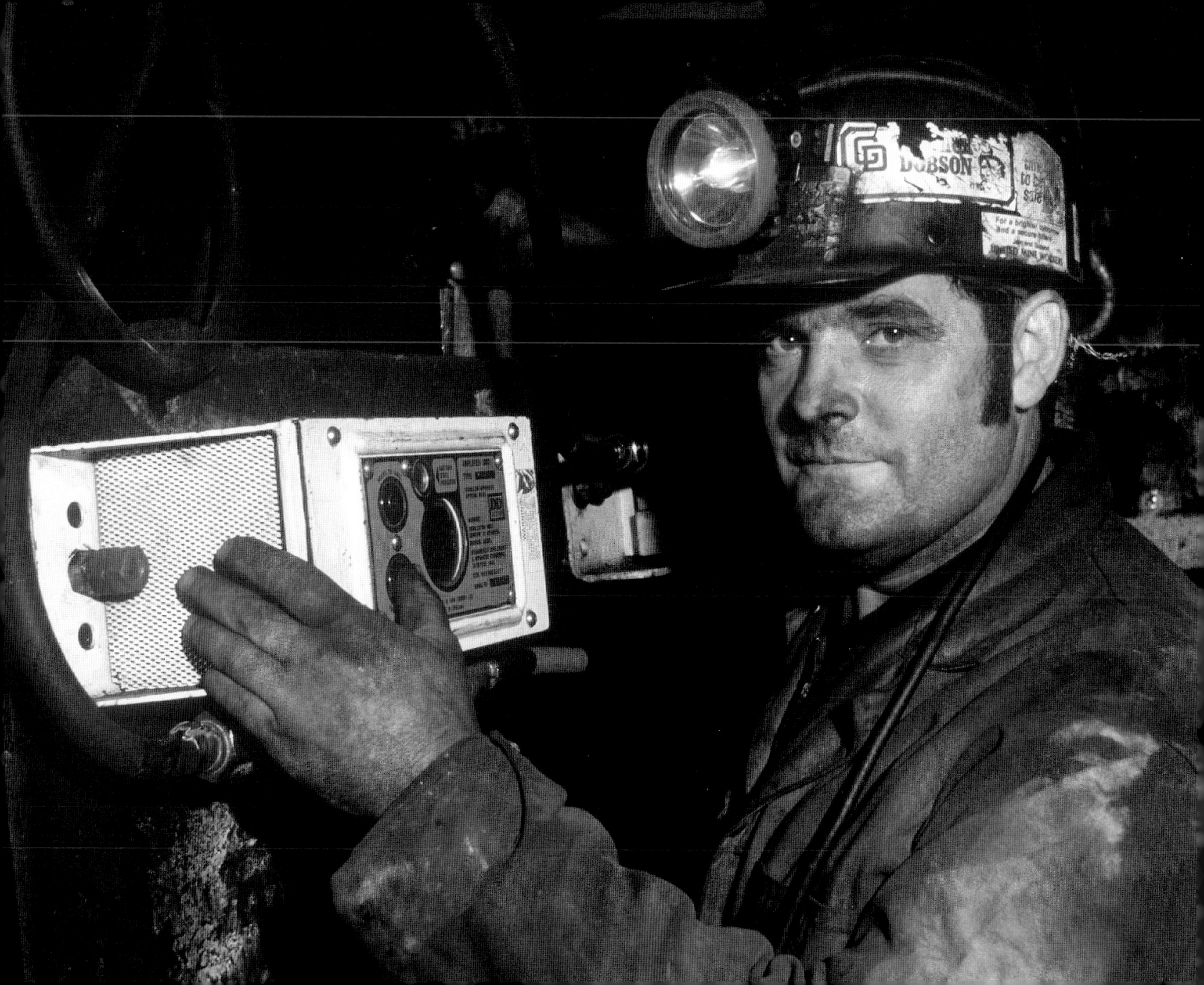
DOBSON

On the left is the haulage chain, next to it the pan
on the right is the shearer.

137-A-10A Lingan Mine 1981

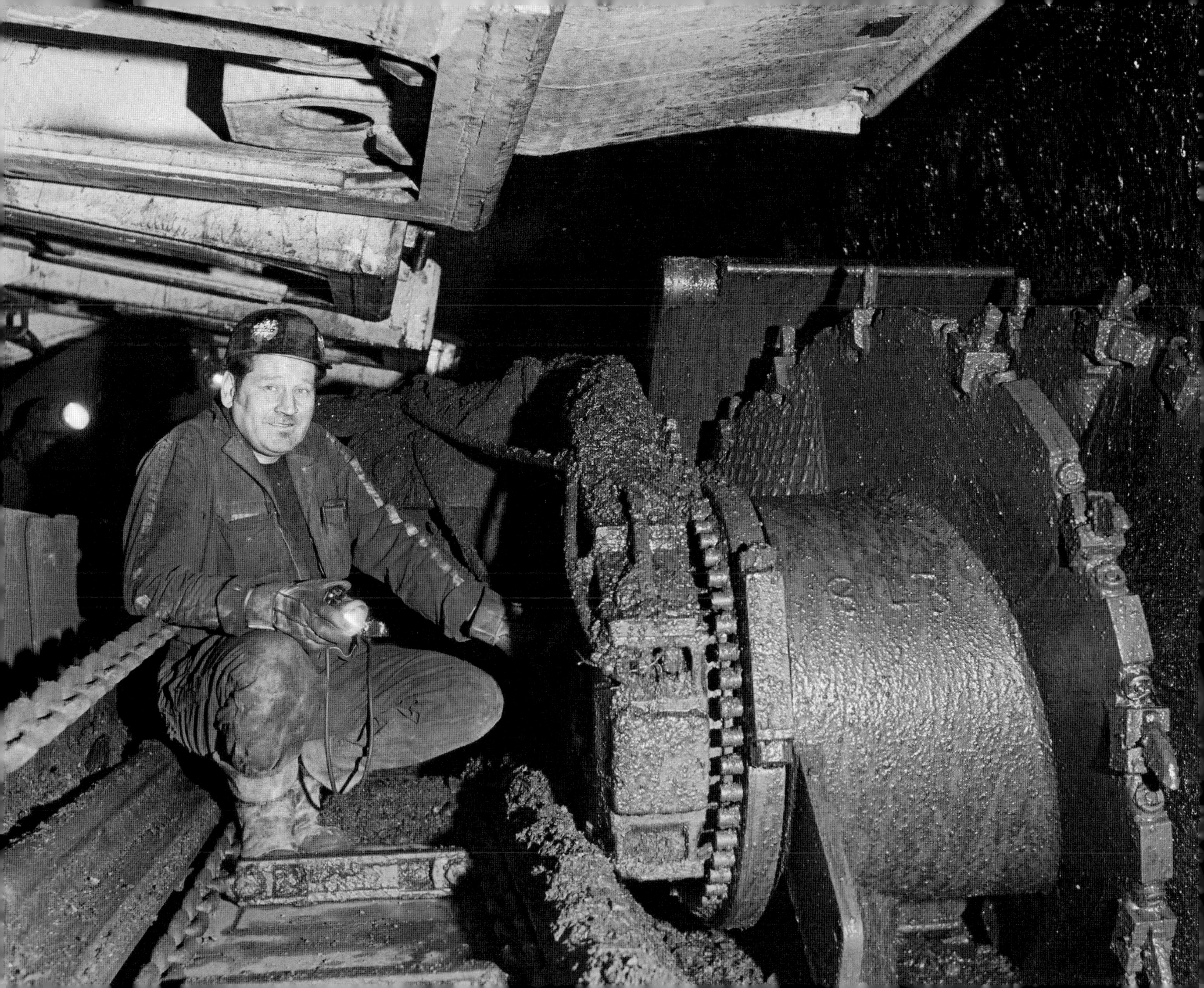

Water was sprayed to keep the dust down as the coal fell on the pan-line.

206-A-28 Lingan Mine 1983

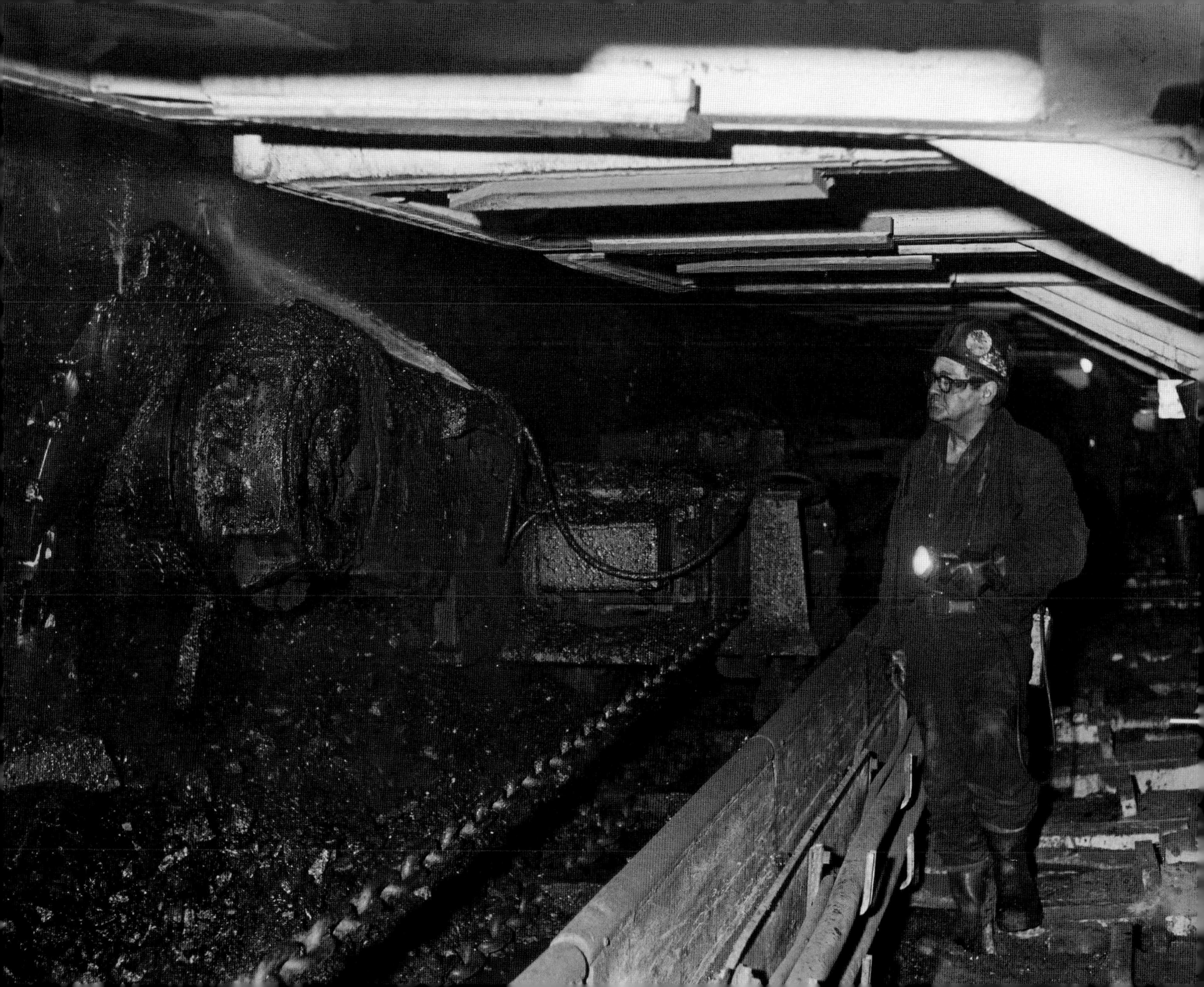

This is an EIMCO – 612 or 627. It had an electric motor. (Combustion engines could not be used in the mine because of the risk of explosion.)

There's barely enough room to get the equipment underground. As soon as new equipment arrived, we started cutting it up. We had to transport it down in pieces, then reassemble.

330-C-7 Lingan Mine 1986

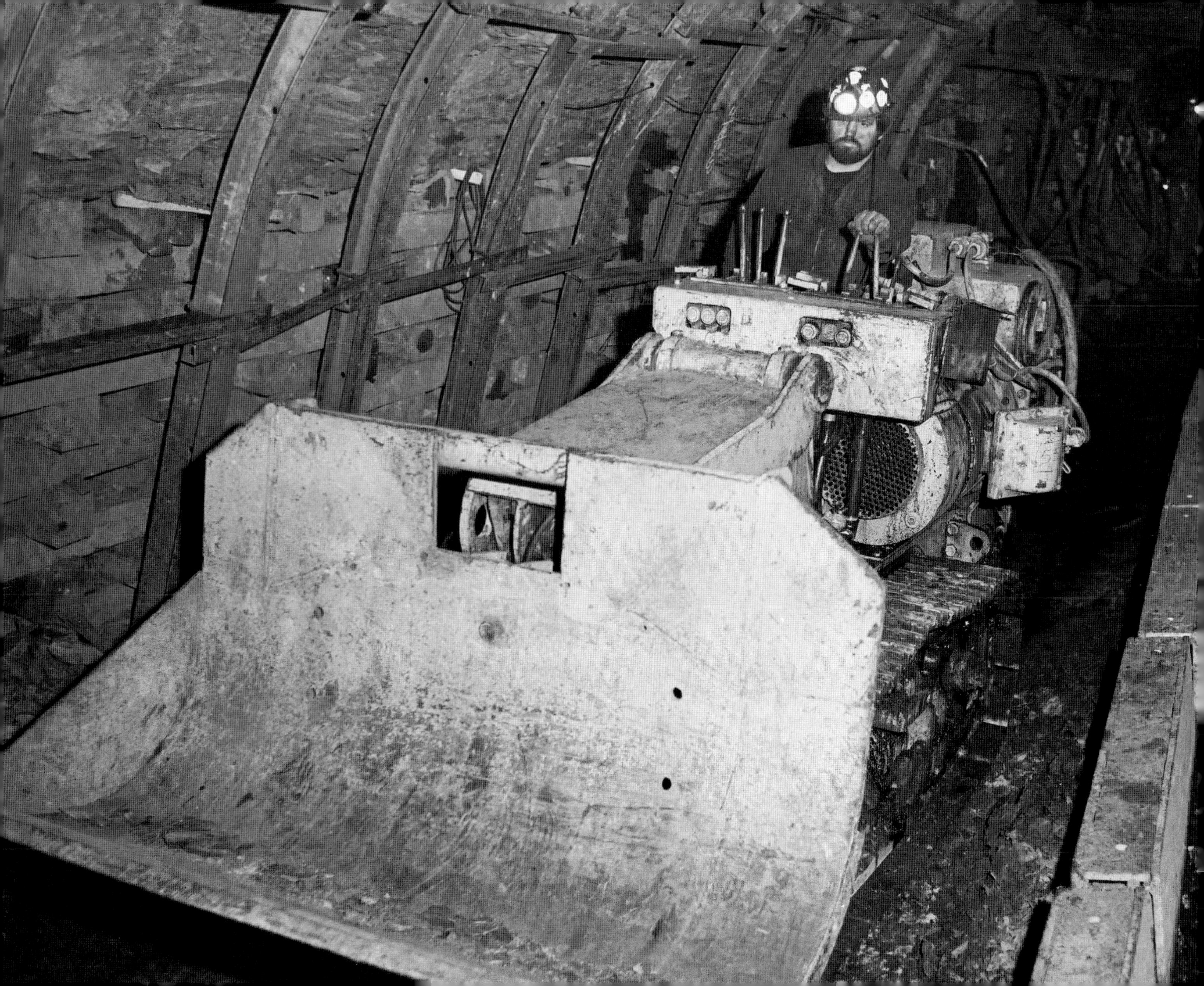

Well, see you'd set your machine up and you'd cut your coal on the face; you set your machine up to the face and you'd cut the coal and then you drill the hole and then you blast it, then you load it. Your partner either was loading it left hand and say I was loading it right hand. Then we'd load the box, push it out to the spare road, take the empty box in, fill it again, all day until the shot of coal was cleaned up.

Tape-2134 23 July 1984.

Holes were drilled for the setting of charges.
The hole they drilled is visible on the wall behind them.

330-C-17 Lingan Mine 1986

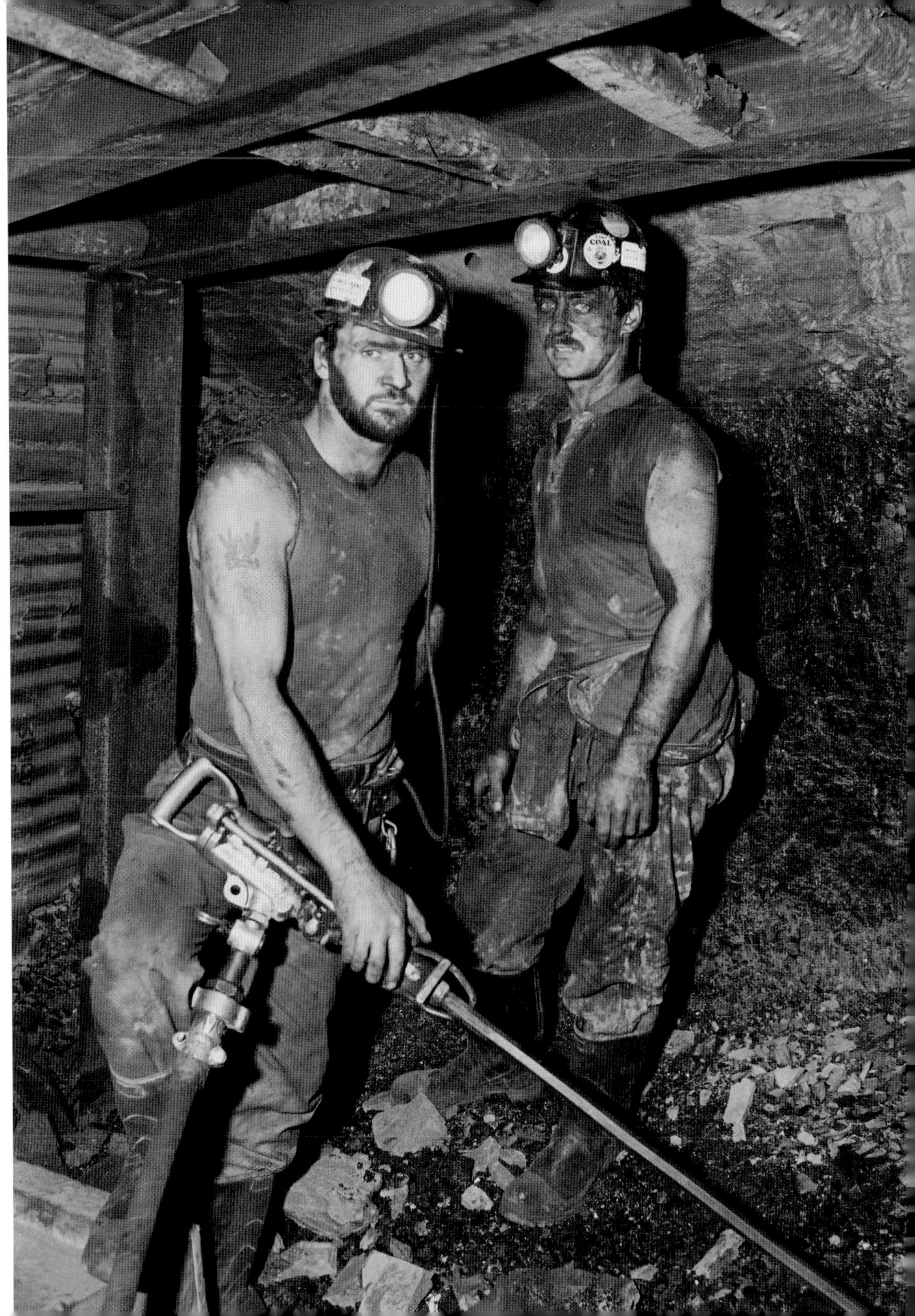
COAL

330-C-26 Lingan Mine 1986

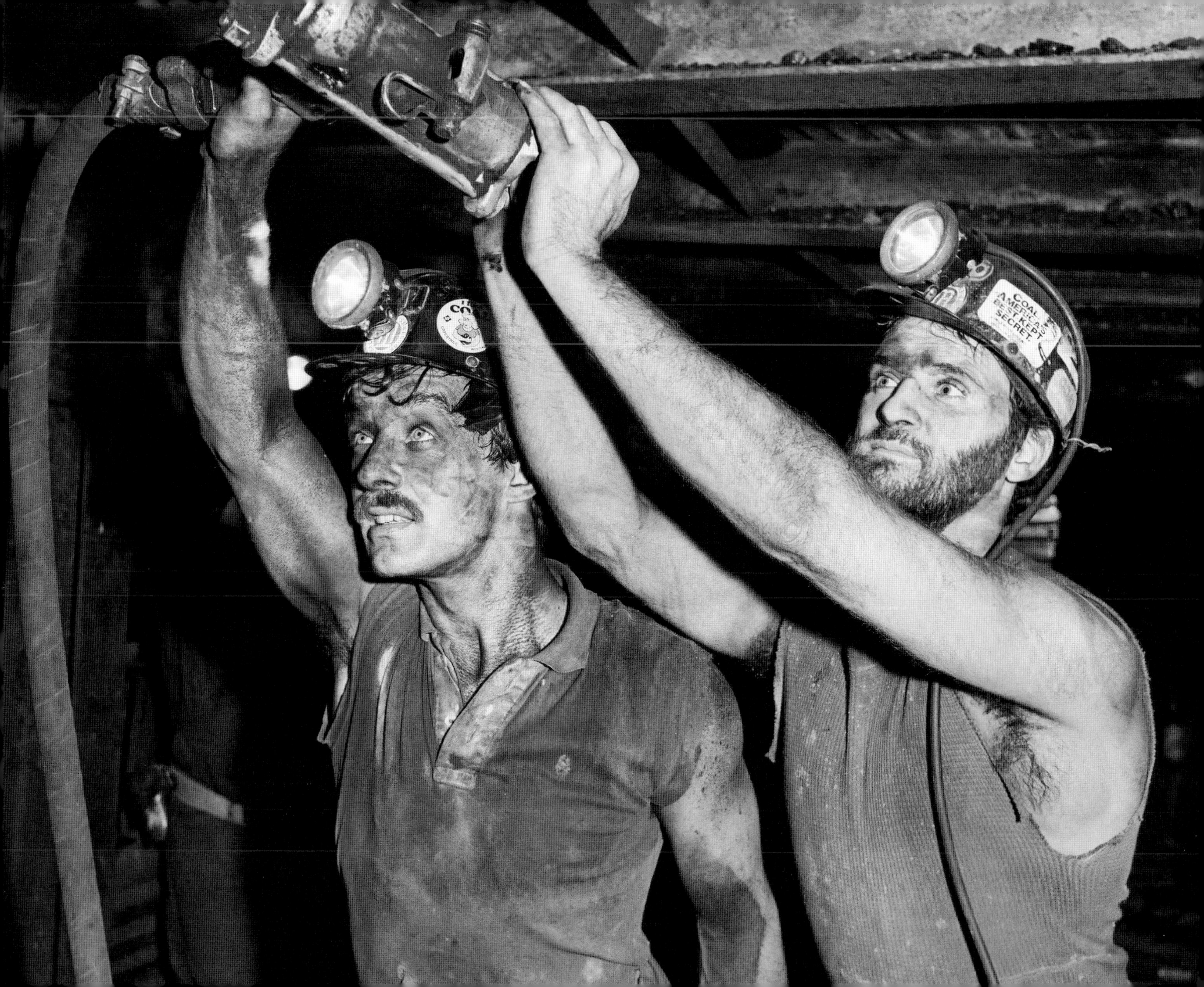
COAL
AMERICA'S
BEST KEPT
SECRET.

Clearing the roof supports before raising them into position.

Tilting, dropping the jacks to take coal from above roof supports.

330-A-24 Lingan Mine 1986

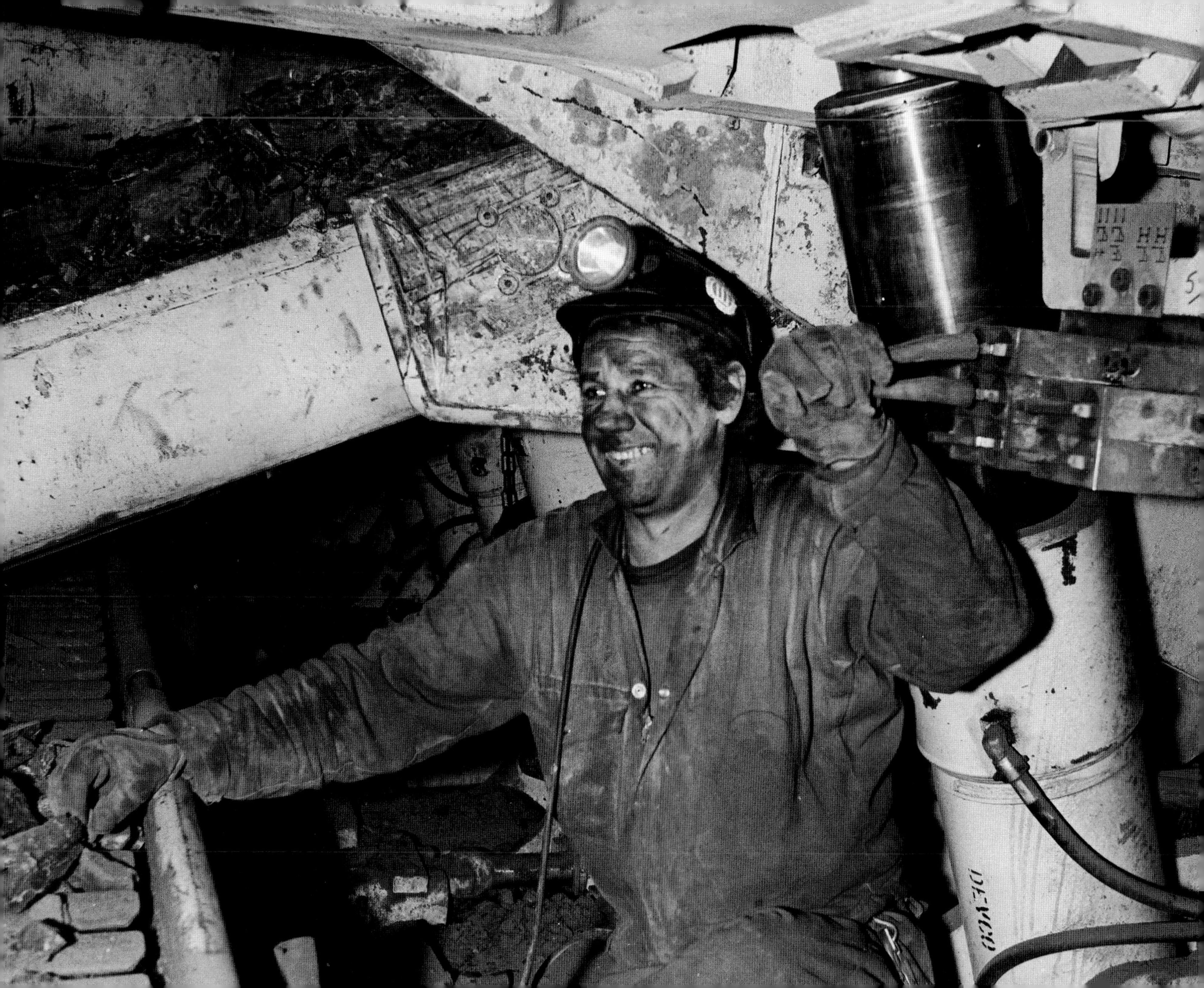

The First Aid teachers saved a lot of eyes
and a lot of fingers too.

Tape-2134 23 July 84.

That's an old First Aid kit on his belt.

330-A-33 Lingan Mine 1986

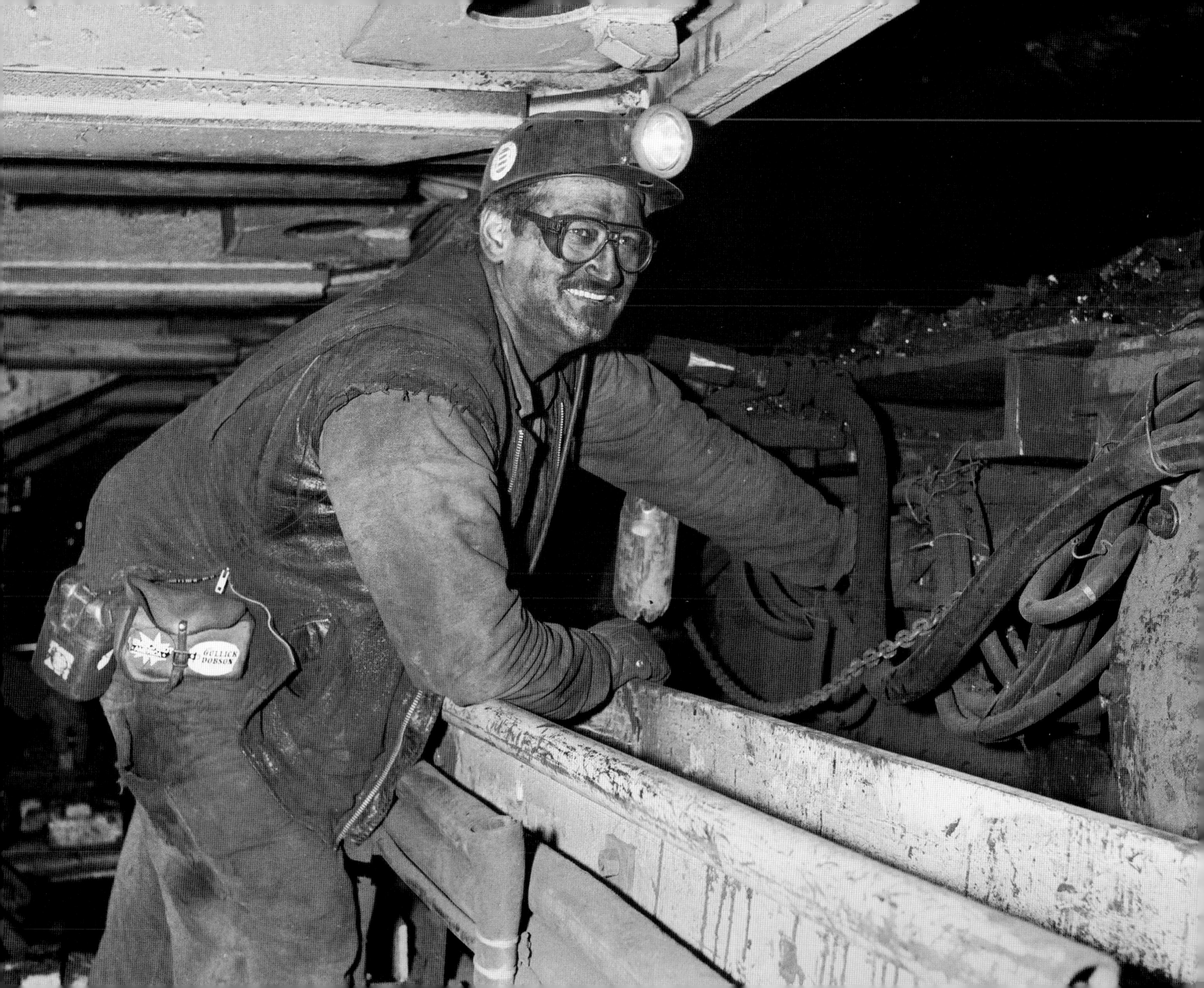
GULLICK
DOBSON

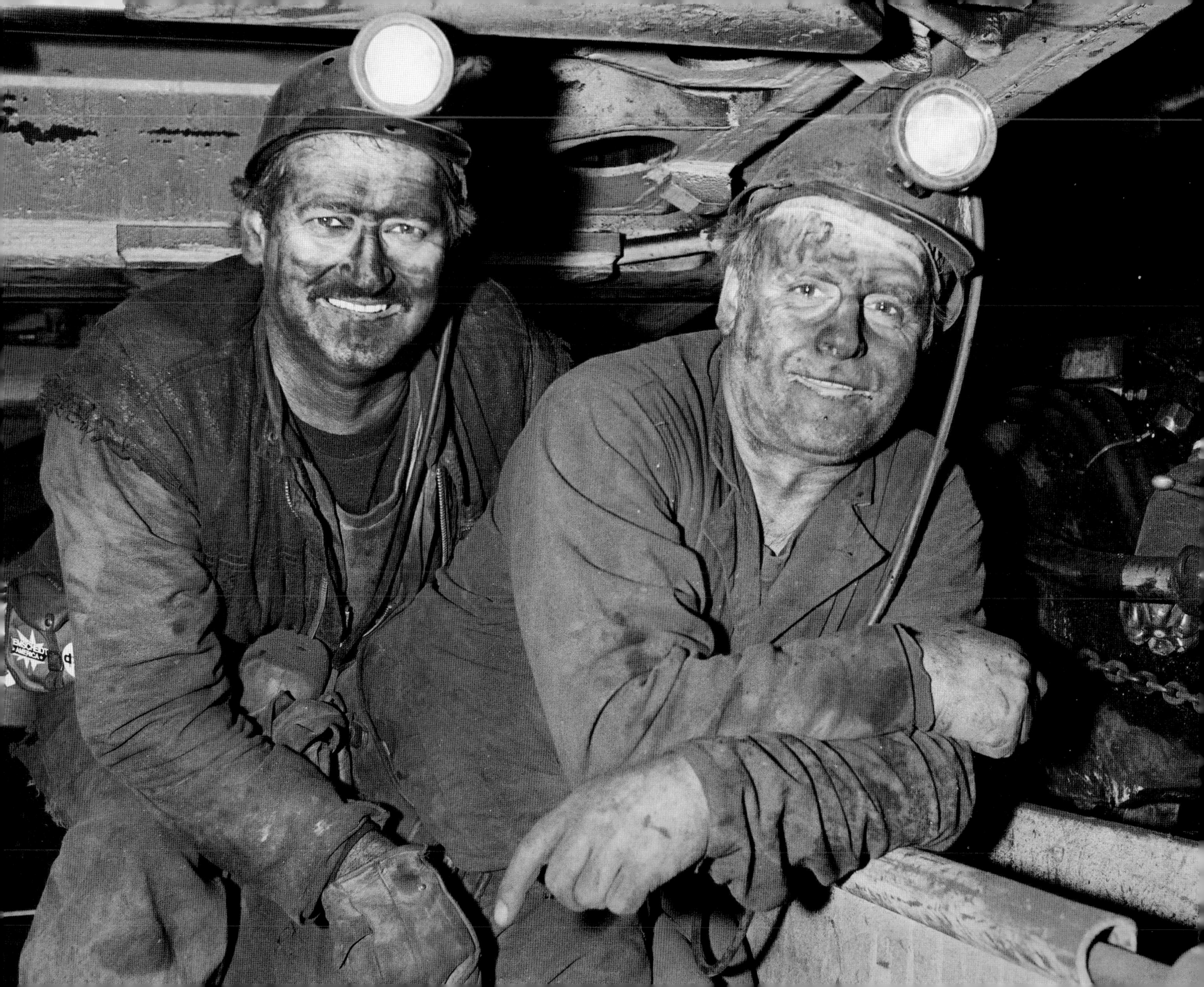

330-A-18 Lingan Mine 1986

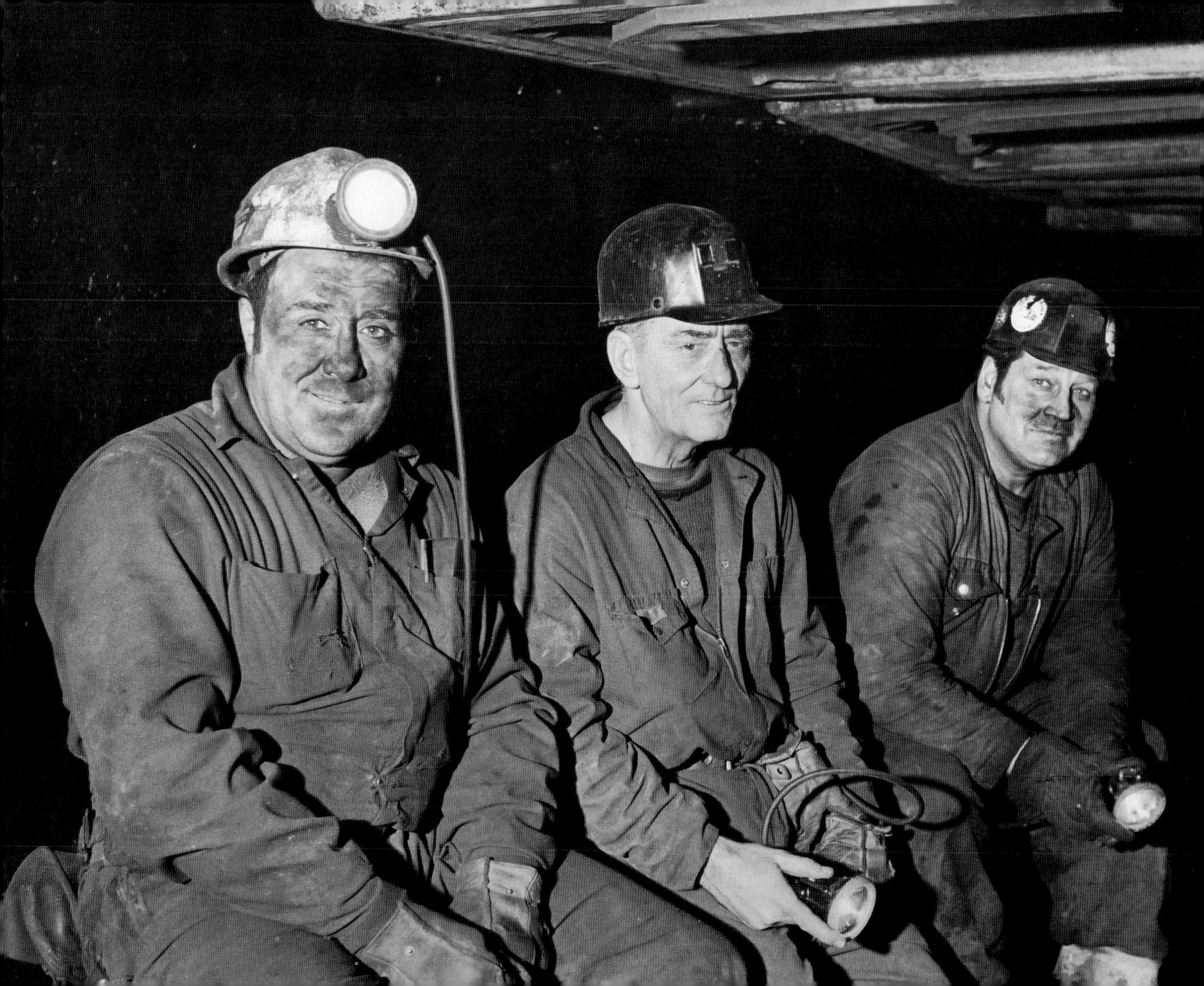

They're on a "landing" waiting for a ride up the deep.

330-B-36A Lingan Mine 1986

Inset 330-B-7 Lingan Mine 1986

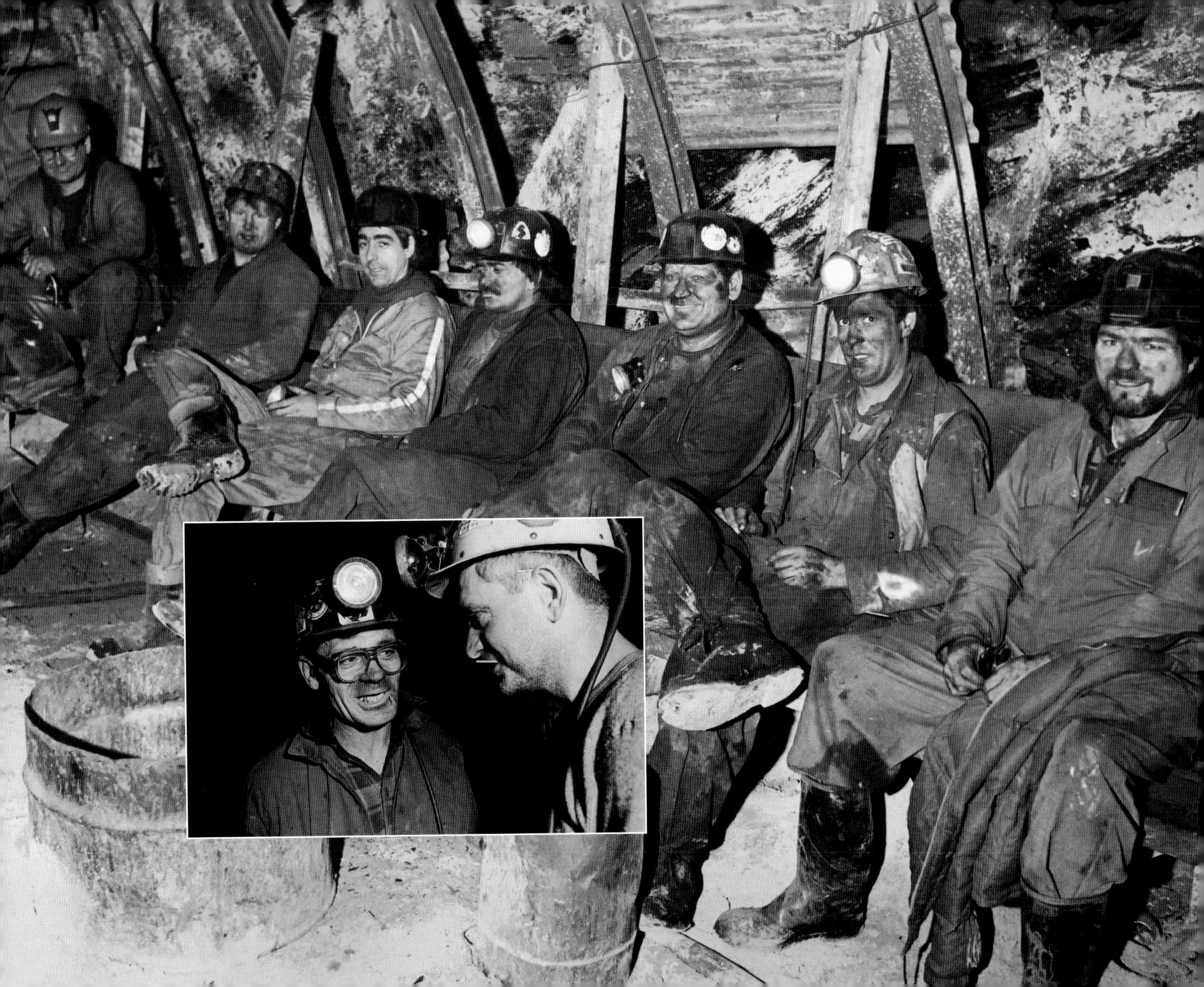

I had to look after a ventilation fan and a man hoist. That shaft was 750 ft. (approx. 225 m.) deep and there was a cage run up and down for hoisting men. The fan went continuously to supply fresh air into the colliery below.

Tape-2146 20 August 1984.

(Right) Fire inspector.

(Left) Belt-operator's shack. He sat there all day to watch the belt. The wind was so strong, due to the ventilation, built a shack to keep warm. He'd freeze if he didn't have that shack. The men used to joke about him reading Old Western novels.

Right 317-A-16 Lingan Mine 1985

Left 302-A-5 Lingan Mine 1985

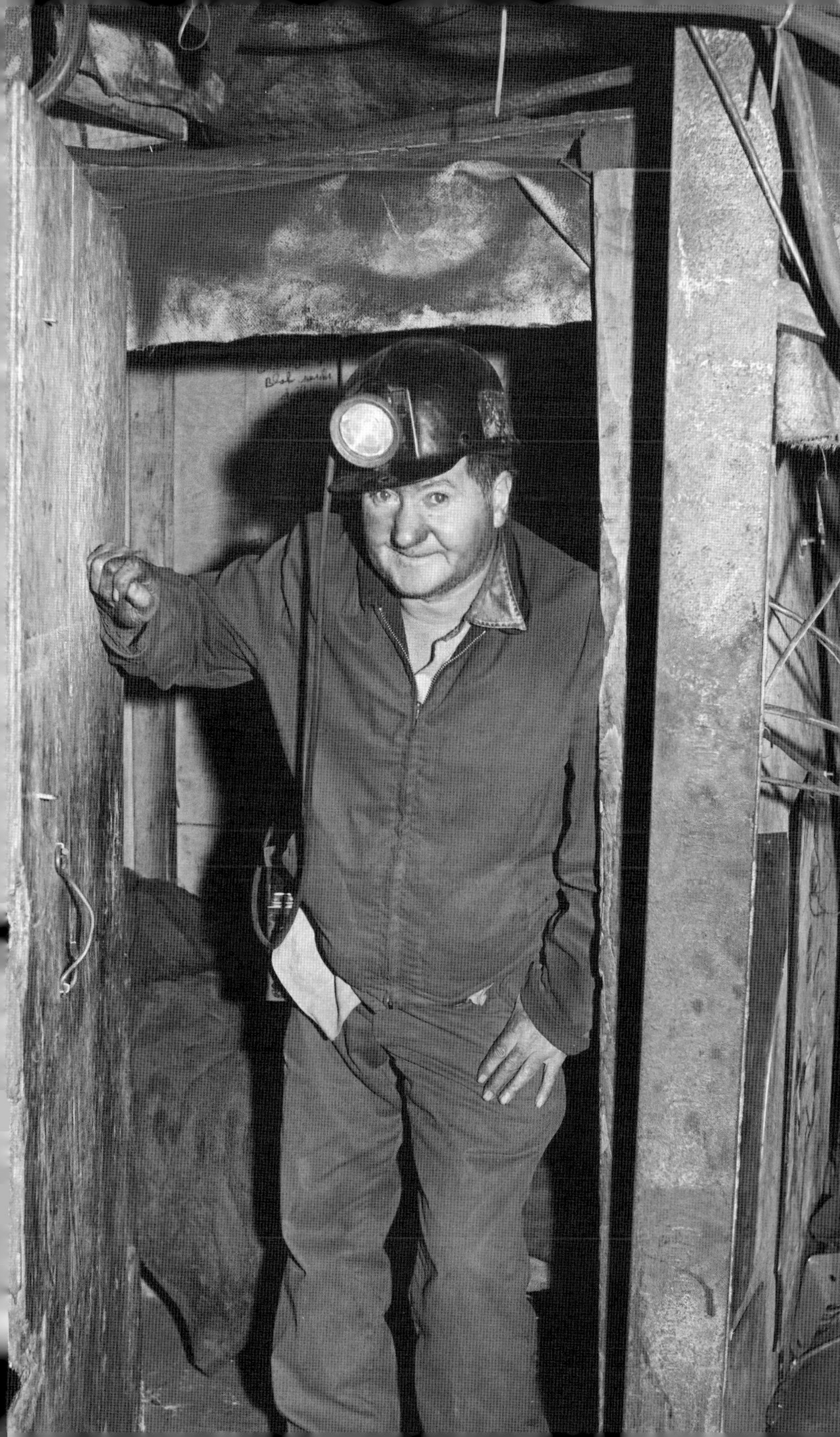
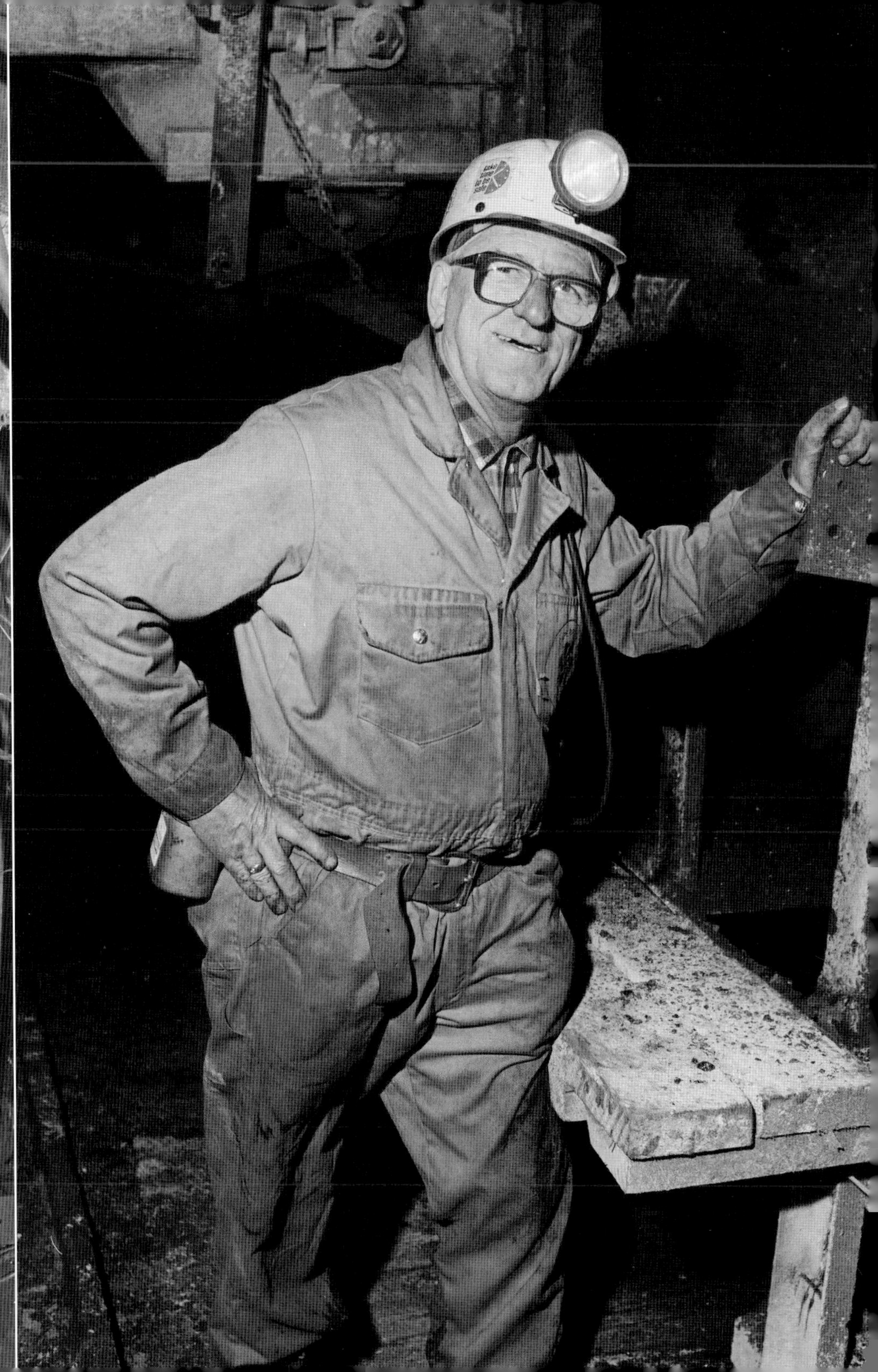

My first trip underground in 1974, was at the Lingan Mine in New Waterford. I was at the mine at 6am to get geared up with boots, under and outer wear, helmet, pit light and the all-important self-rescuer. The self-rescuer is a breathing apparatus that could give you an extra 90 minutes to escape incase of fire or a cave-in.

The coal workings were continually moving further away from the shaft and a great deal of time was consumed travelling to and from the working face. Princess colliery claimed that its underground railway was one of the world's longest underground transportation systems. The men were taken from the shaft bottom to the mine face aboard a man-rake, small flat rail cars pulled by a cable. These man rakes had no top or sides. In newer mines, the man rake was completely enclosed. The distance from pit bottom to the last long-wall, No. 27 south, was nearly seven miles.

I got on the man-rake, which consists of a series of small cable cars, which lower men and supplies into the mine. Then, for one hour, the rake raced down the "deep," which was the main tunnel into the mine.

It was pitch black, except for the pit light lighting the low ceiling overhead. Every hundred feet or so, along the deep, there were small dug-out areas in the wall, called "manholes." These were big enough for one or two men to duck into, in case of a runaway cable car. During an underground inspection in the 1980's, Steve Rankin, then president of DEVCO, was severely injured by a runaway man-rake.

Once we got to the pit bottom, we left the rake and walked about half-a-mile into the actual coal face where the coal is mined.

I remember travelling the old Princess Mine in North Sydney back in 1975. First you were lowered into the bowels of the earth in a cage, then you had to lay back on the man-rake as this cable car speed down the pitch black slope for a couple of miles.

Owen

Open-rake boxes. They're down over a mile—you can tell because of the numbers written on the metal arches.

317-B-9 Lingan Mine 1985

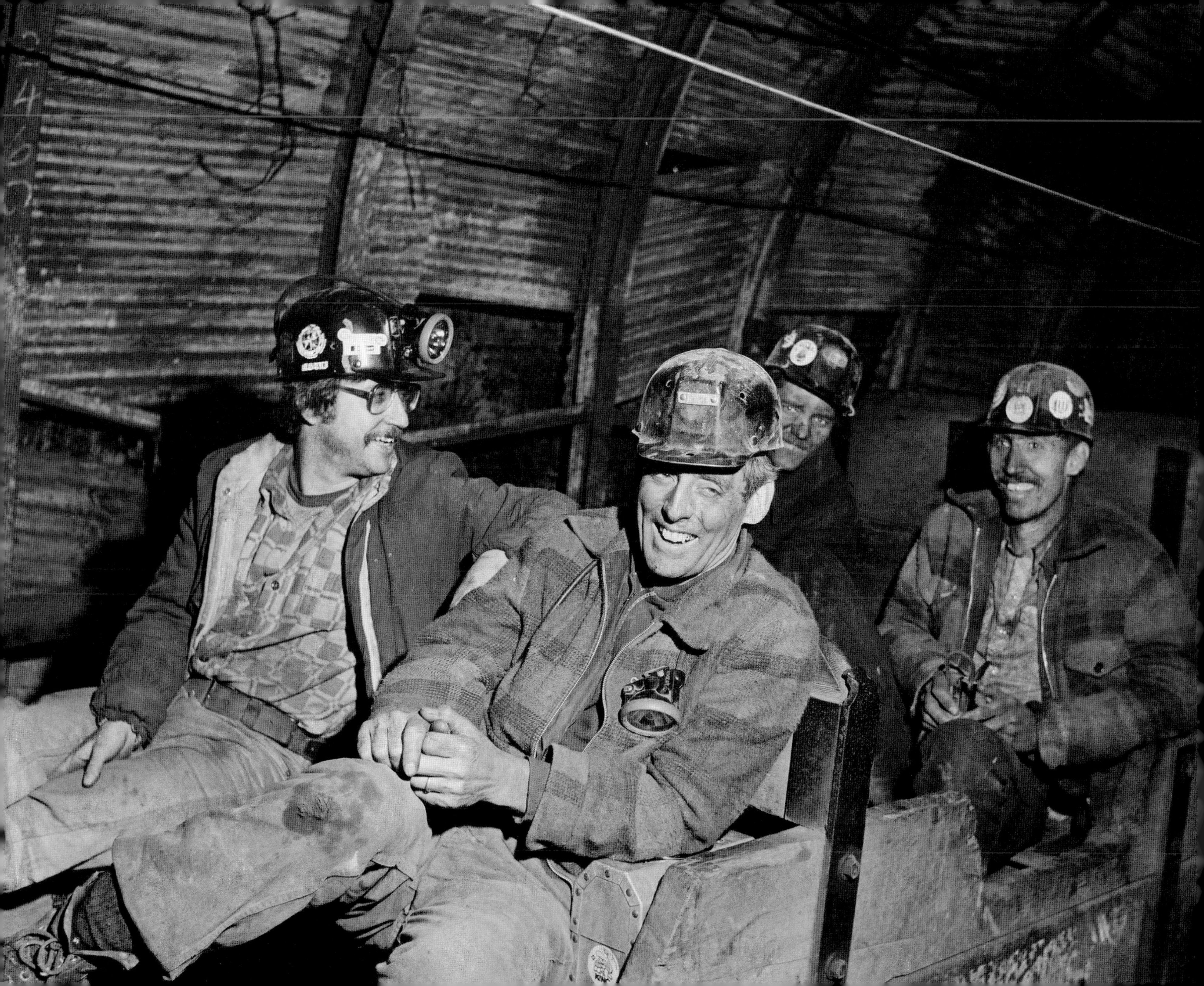

Haulage engine.

330-B-9 Lingan Mine 1986

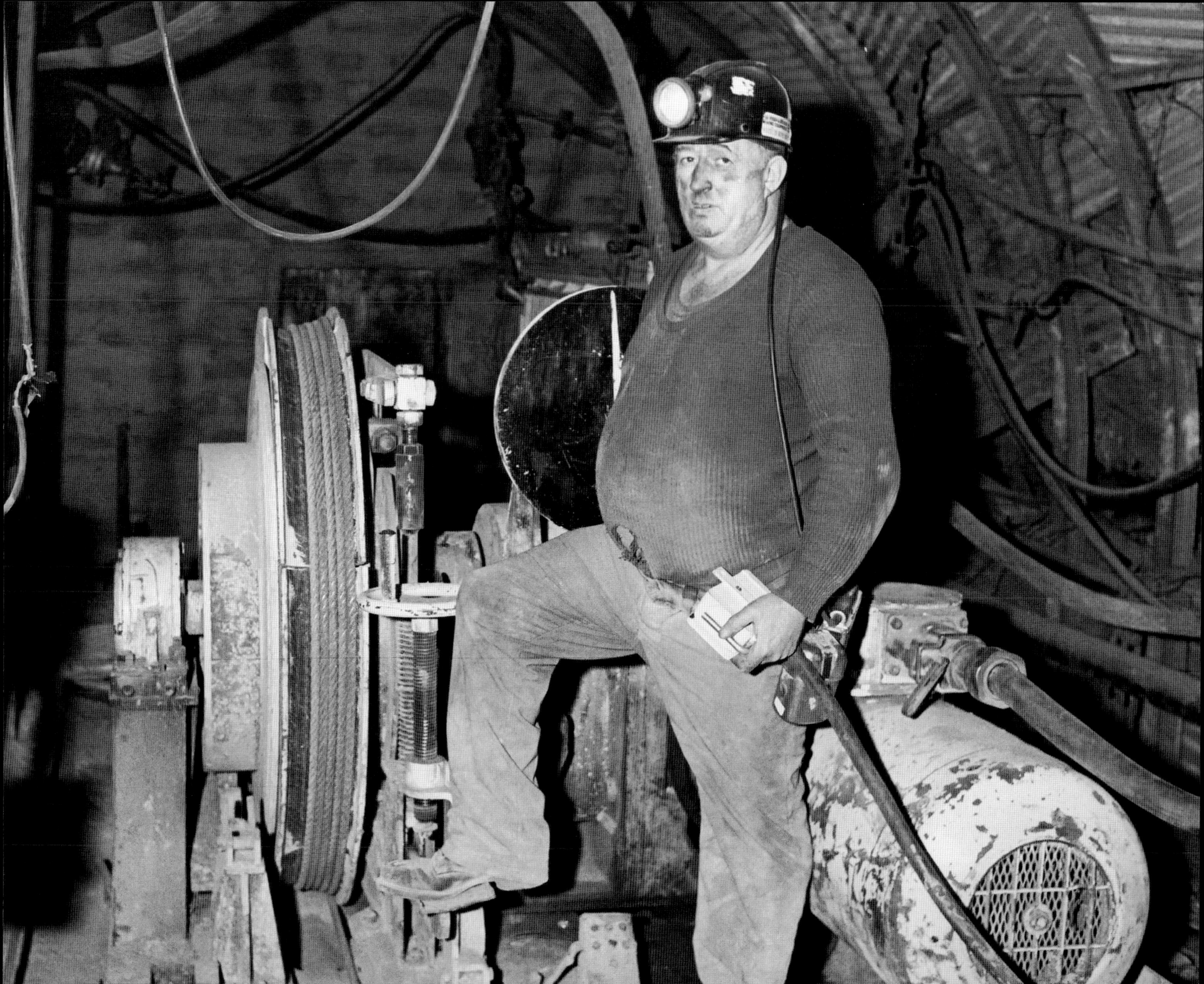

286-A-9 Prince Mine 1980

PRINCE MINE
1980
1986
1987

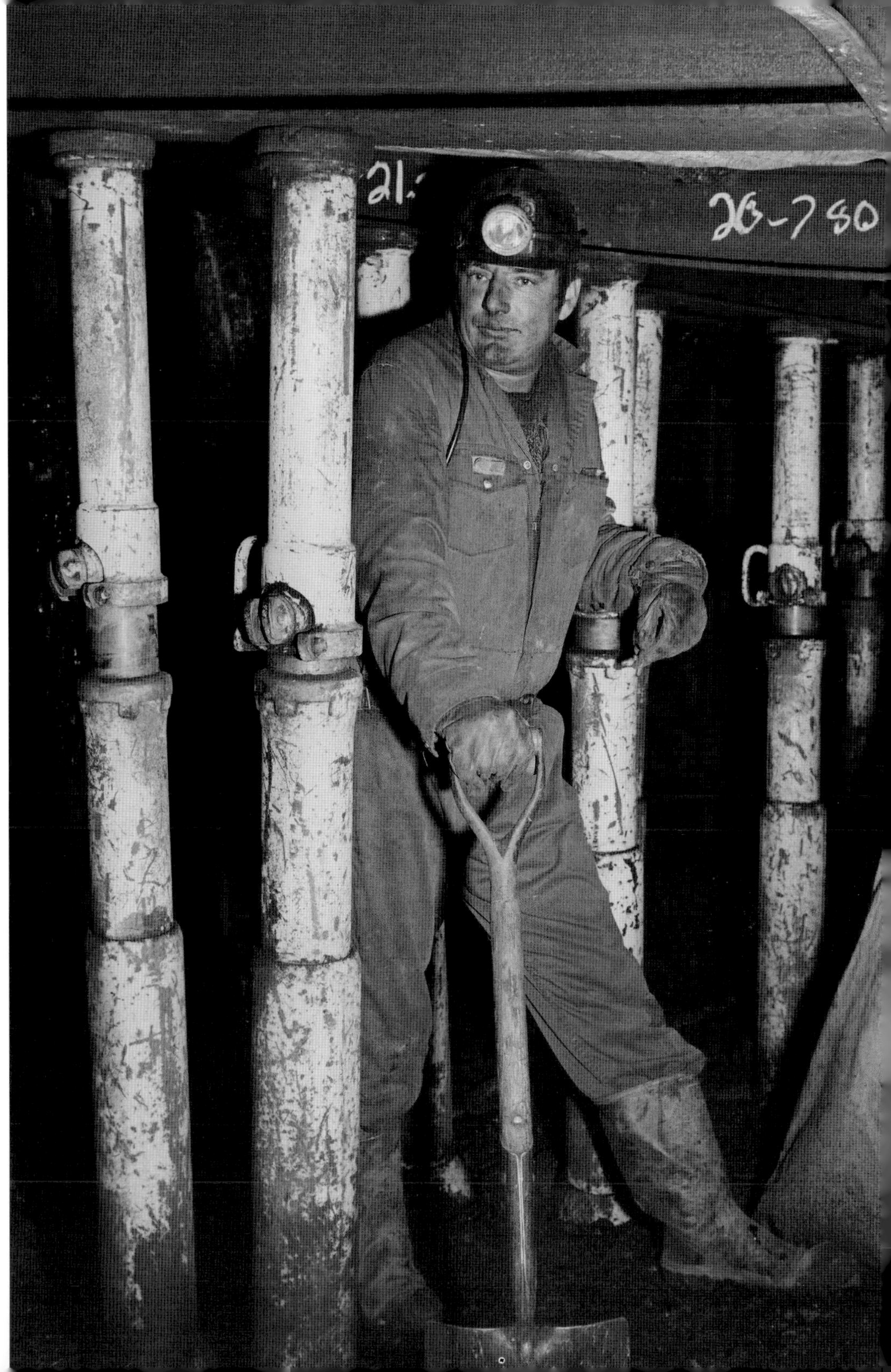

Number 4 slope going down the deep.

361-D-9 Prince Mine 1987

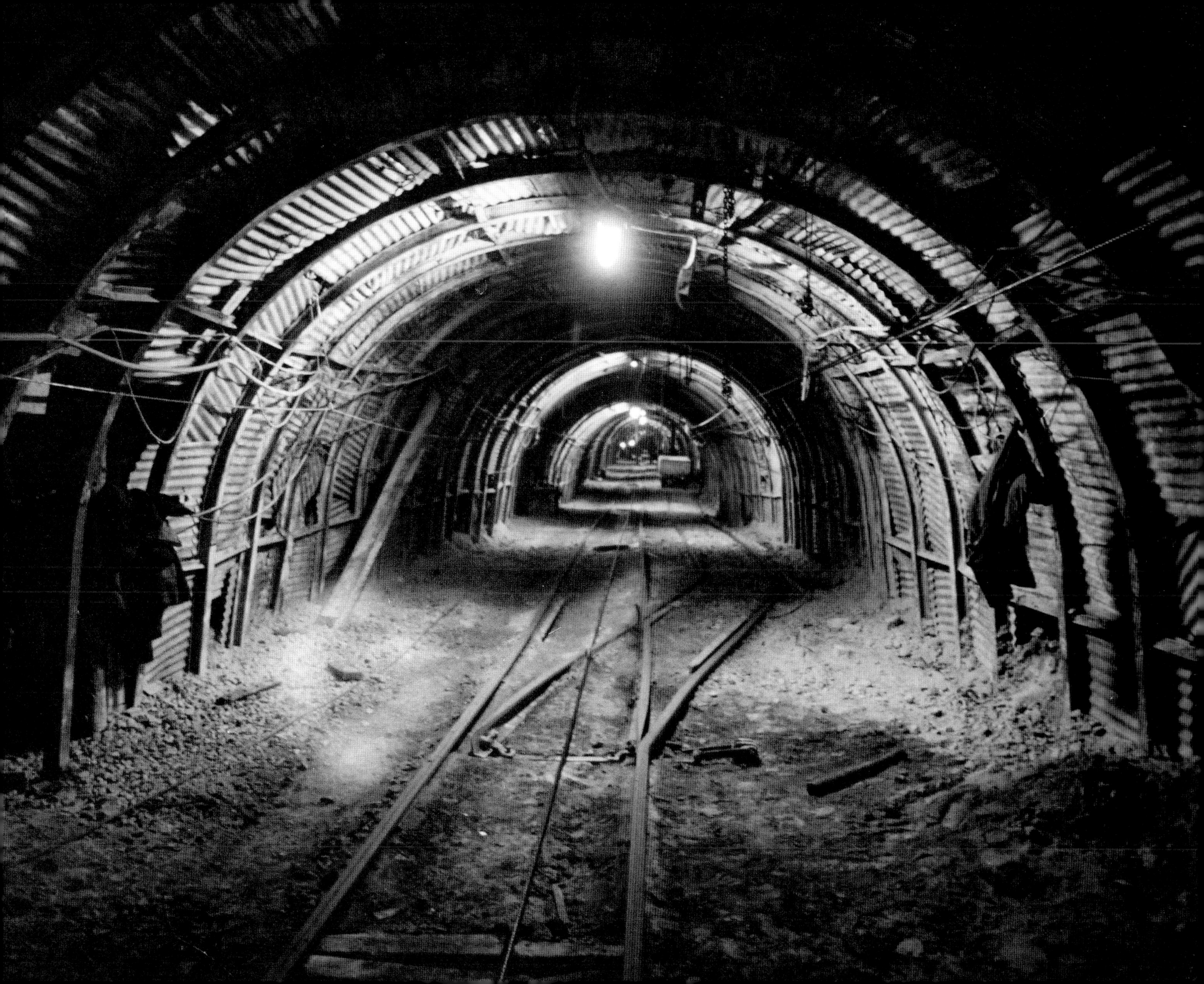

Our garbage cans were just empty oil cans. Garbage cans stood in as toilets, too.

361-C-4 Prince Mine 1987

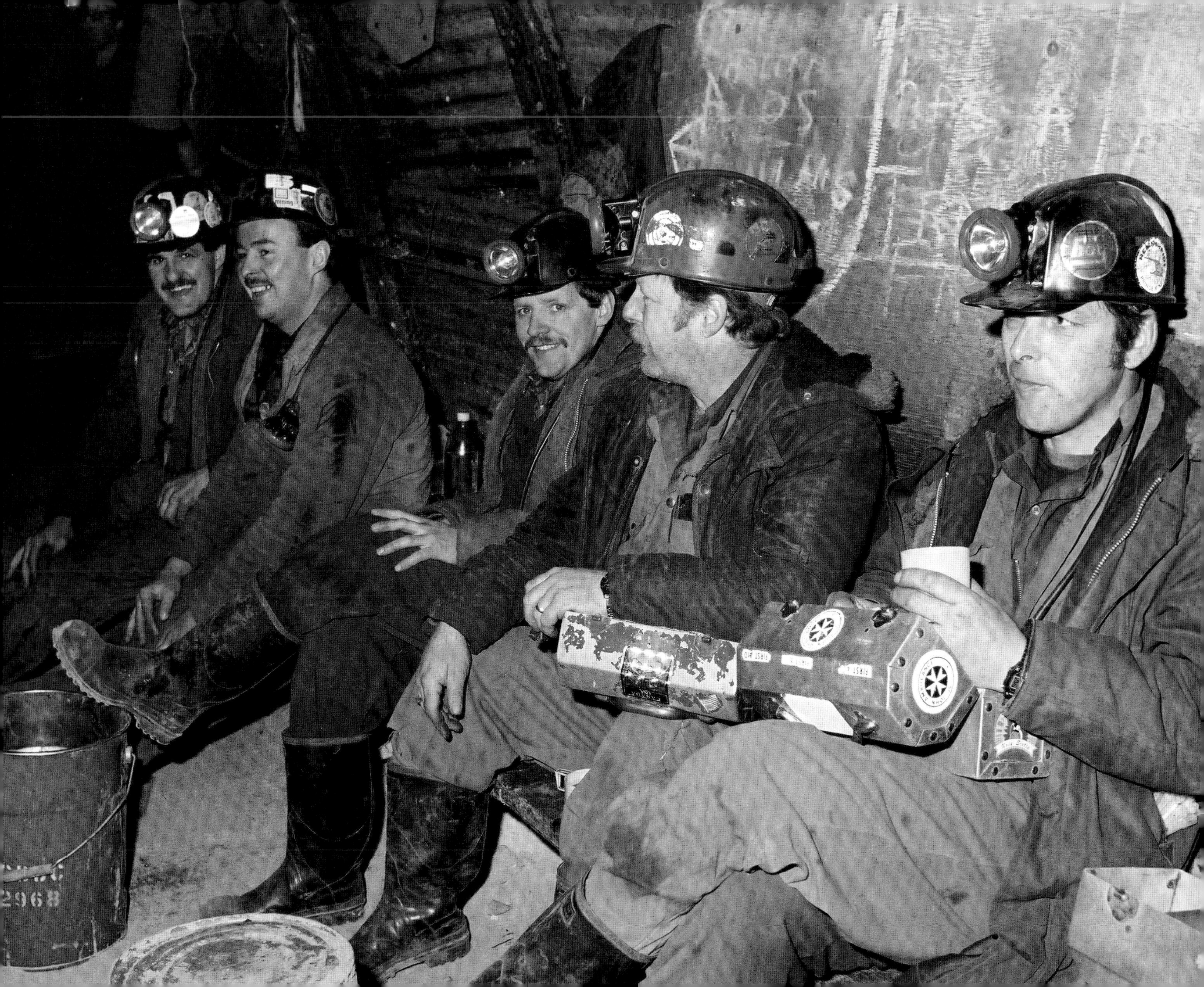
mining
2968

He must have just started in the pit; look at how shiny that hat was! He's from "the other side" [of Sydney Harbour]. Miners were sometimes known by their phone exchange. Glace Bay were 849ers.

286-A-12 Prince Mine 1980

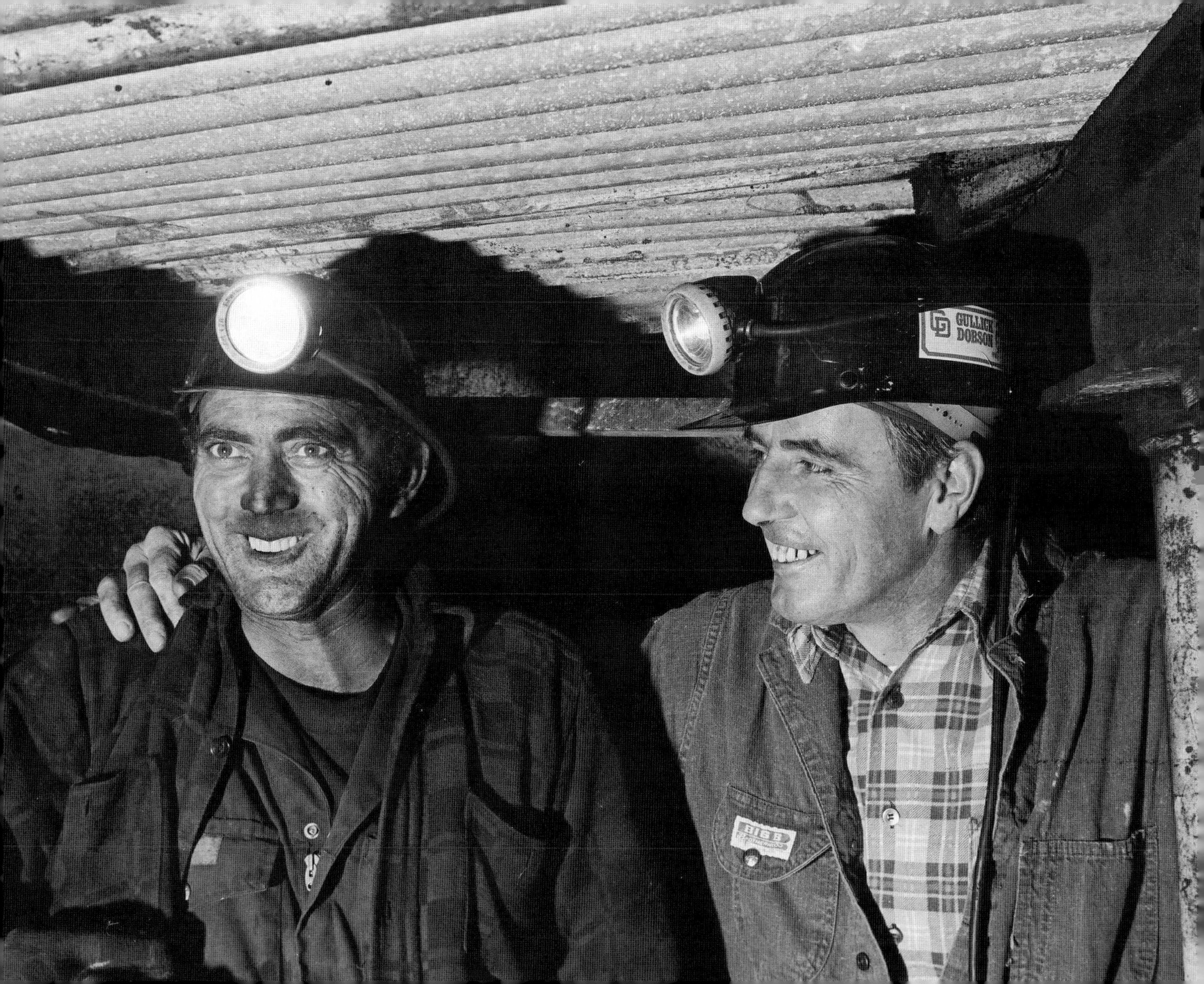
GULLICK DOBSON
BIG B

Used metal tins to support the roof between the booms.

320-A-7 Prince Mine 1986

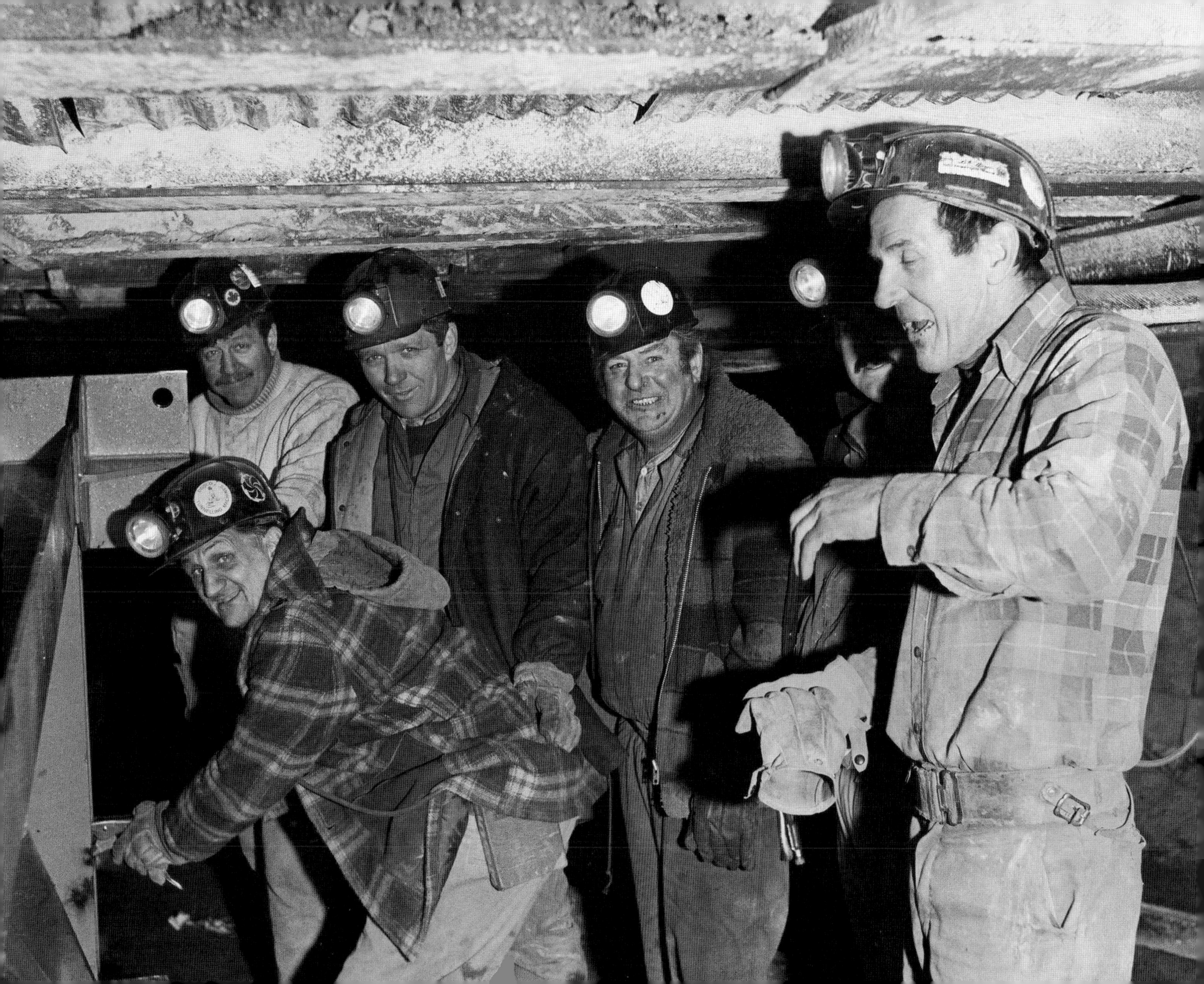

We all got along, everyone got along.

320-A-12 Prince Mine 1986

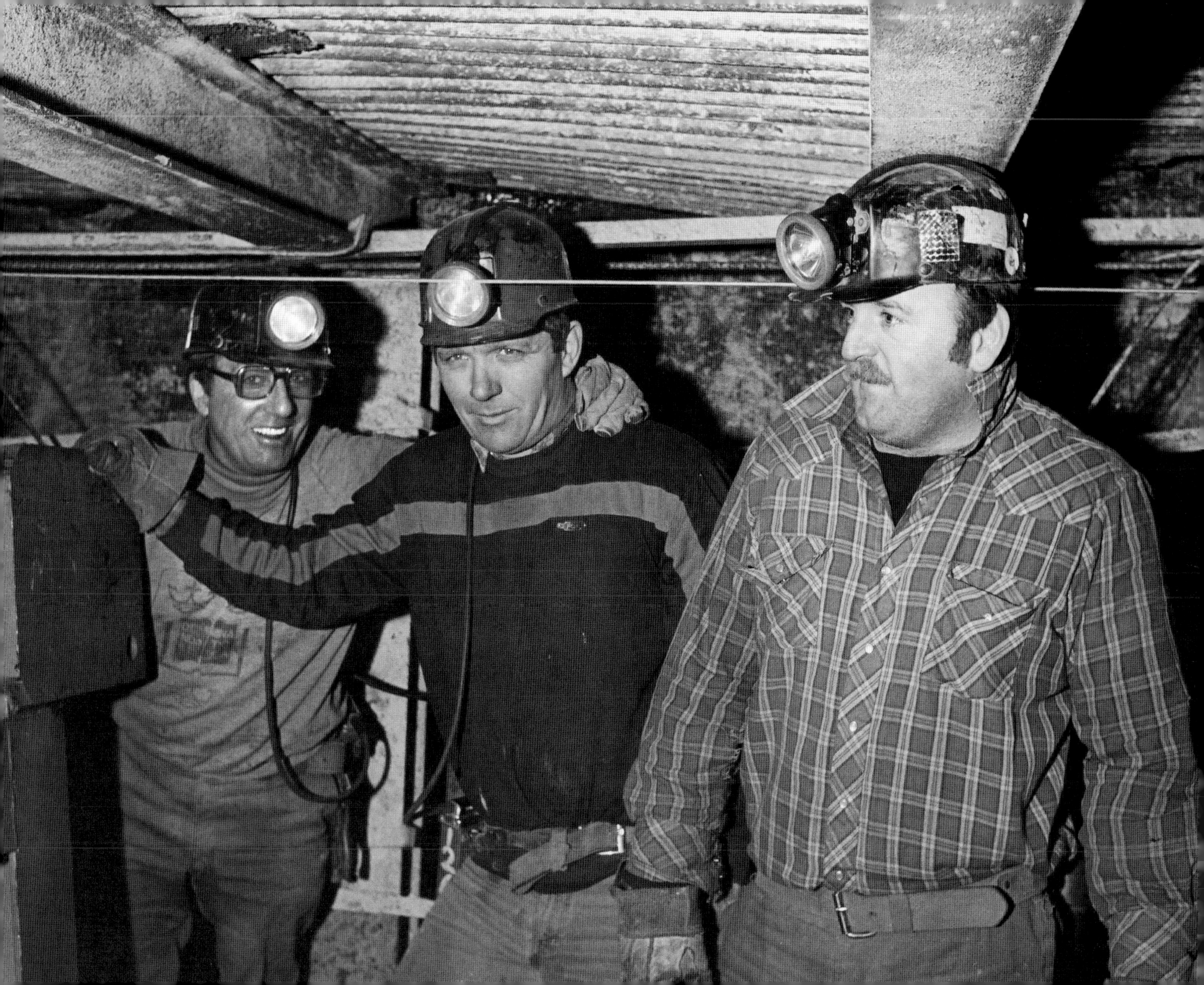

Each mine had different cutting equipment as the seams were different. Each mine had different names for pieces of equipment too. This is a DOSCO dintheader. The chain rotated and the coal fell through the chains.

320-B-7 Prince Mine 1986

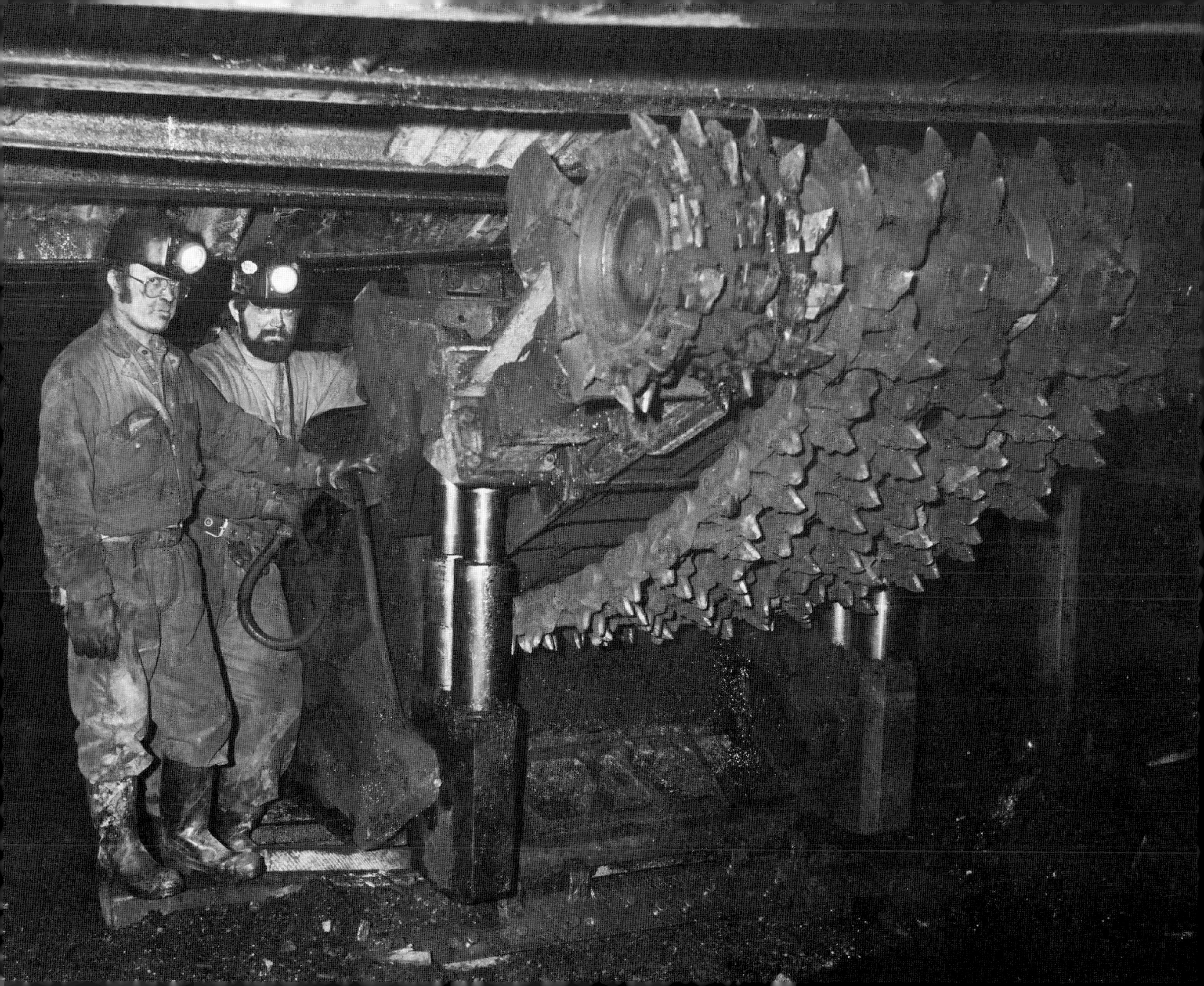

He has a respirator, but he wasn't wearing it over his mouth. You can't spit with it on.

320-C-21 Prince Mine 1986

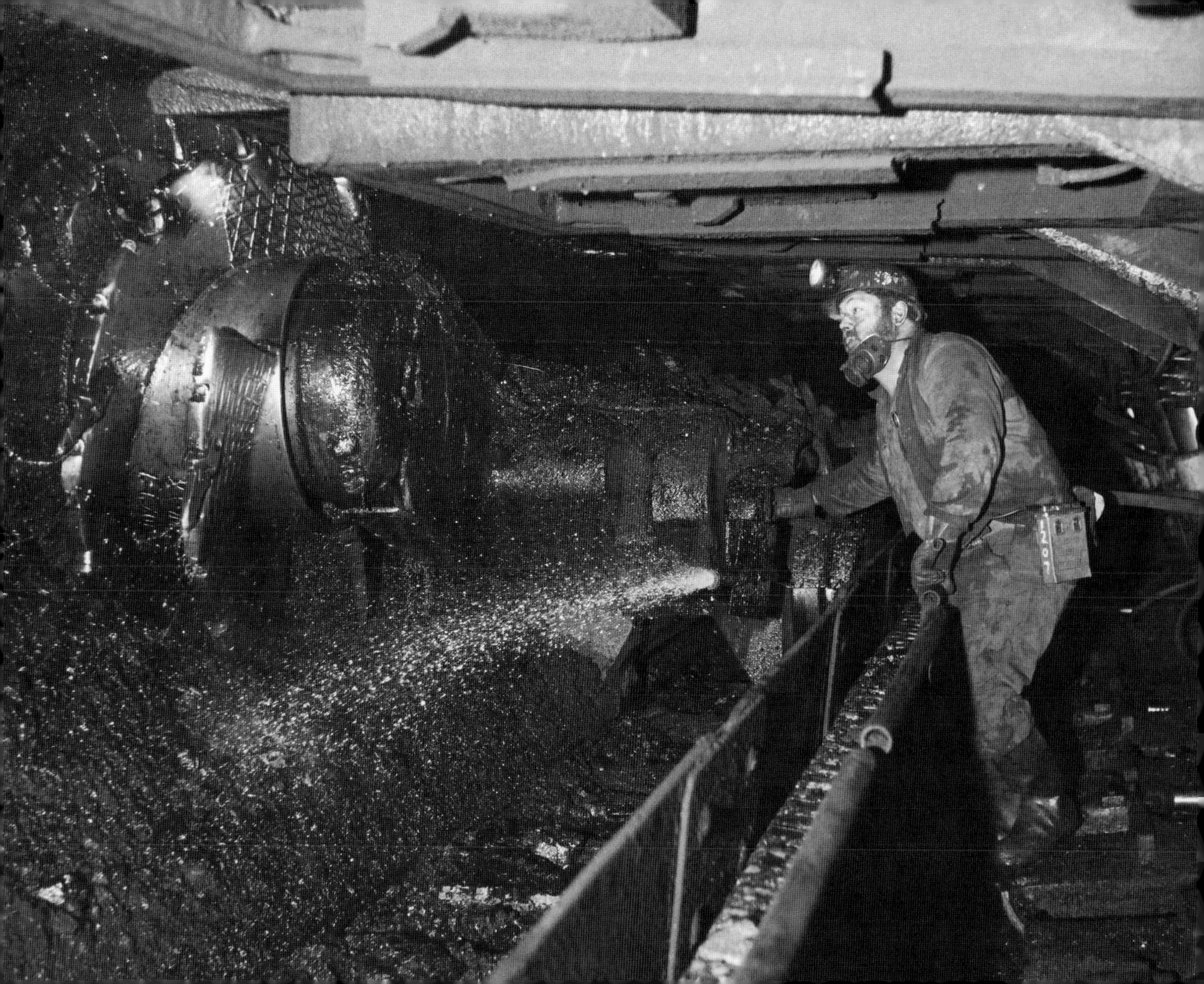

Working at top of the wall. That's a water spray nozzle hanging from top.

320-A-15 Prince Mine 1986

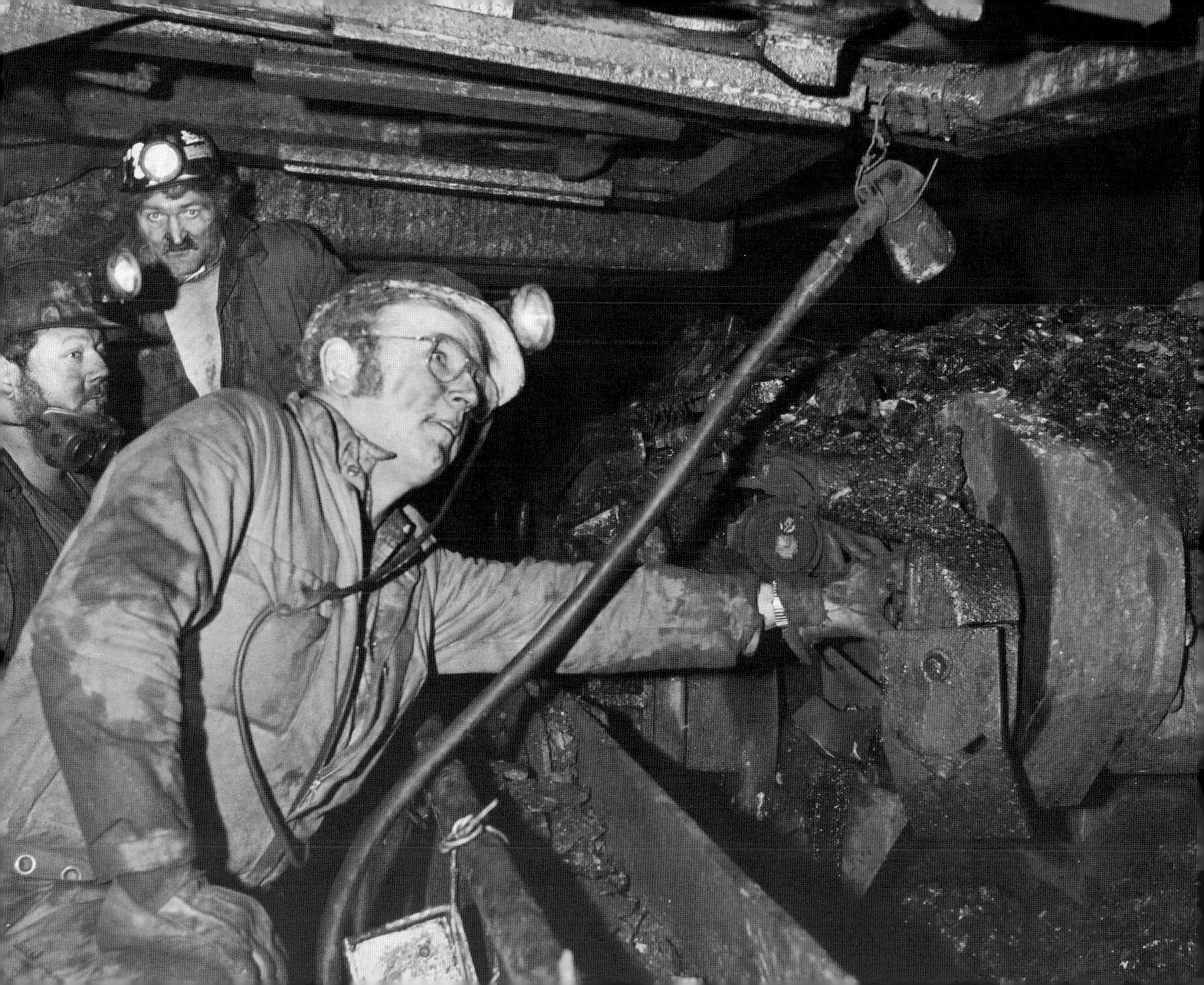

I like it very much. I was doing heavy work, loading coal, but I always loved hard work. My reason for liking it was, in comparison with other work the money was good then and you could make good money if you work hard enough. I didn't mind the hard work at all.

Tape-2157 23 August 1984.

The maul gang worked at breaking the stone.

320-B-29 Prince Mine 1986

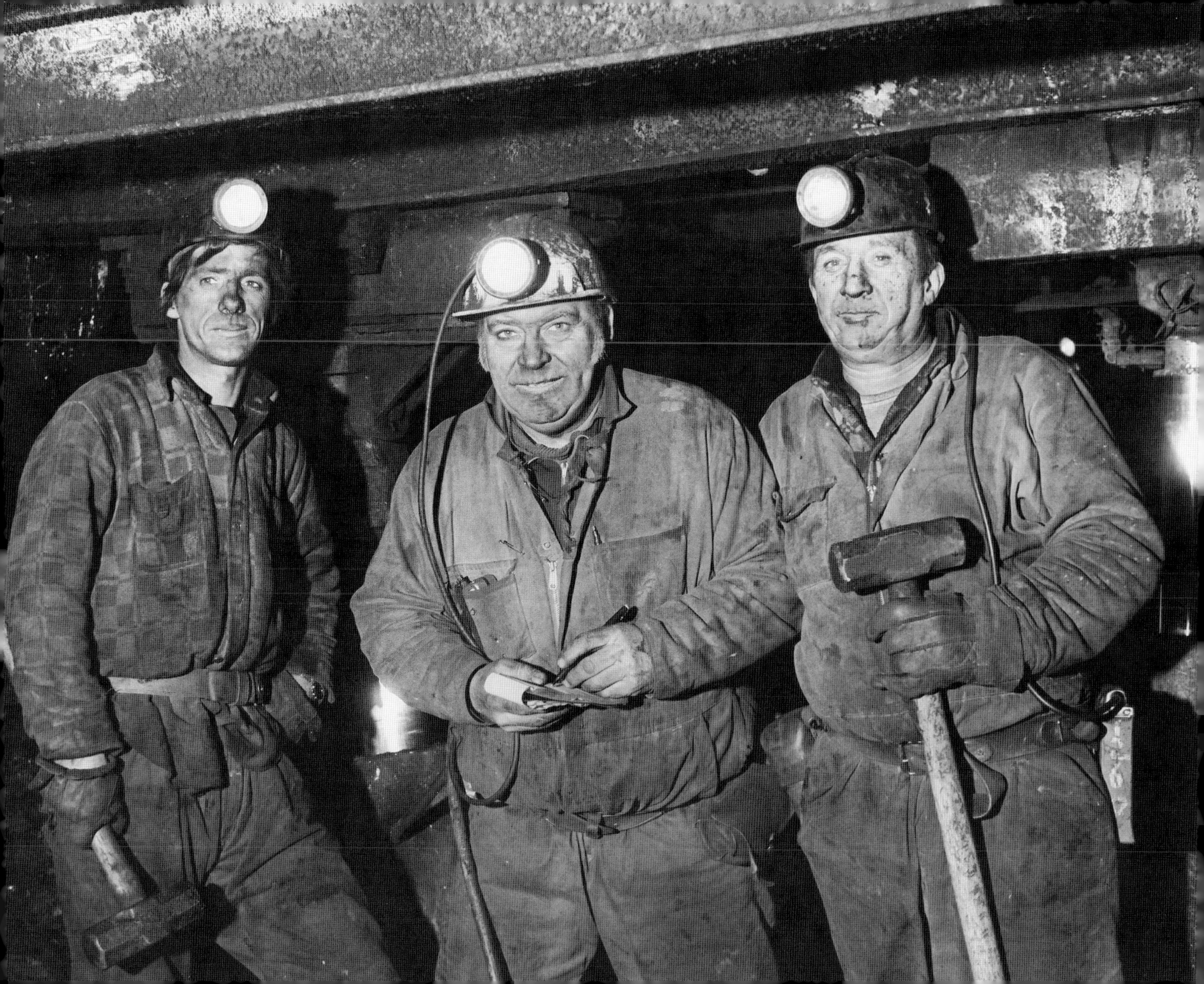

Working with Dowty hydraulic roof props.
You propped them up by hand.

286-B-5 Prince Mine 1980

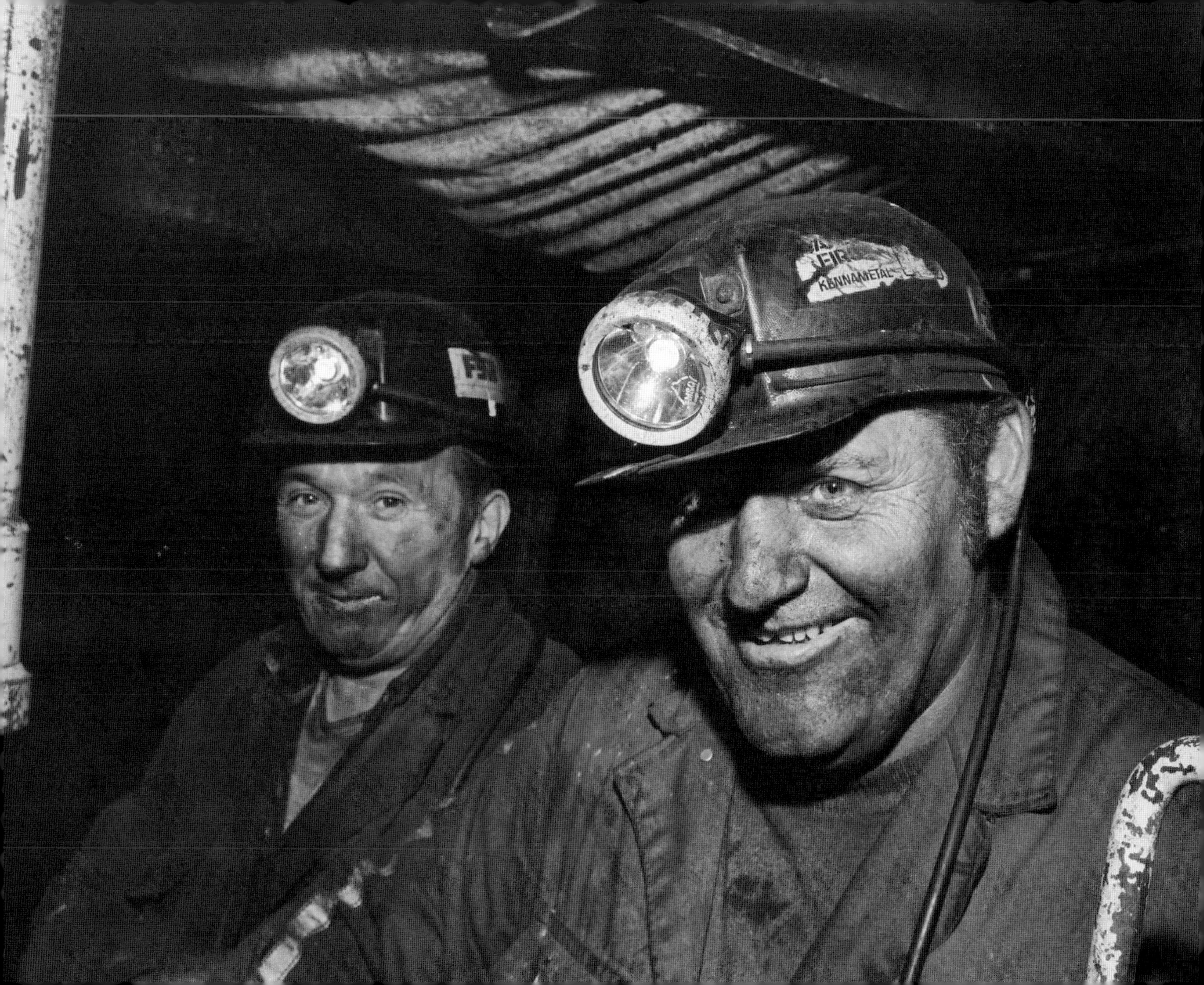
KENNAMETAL

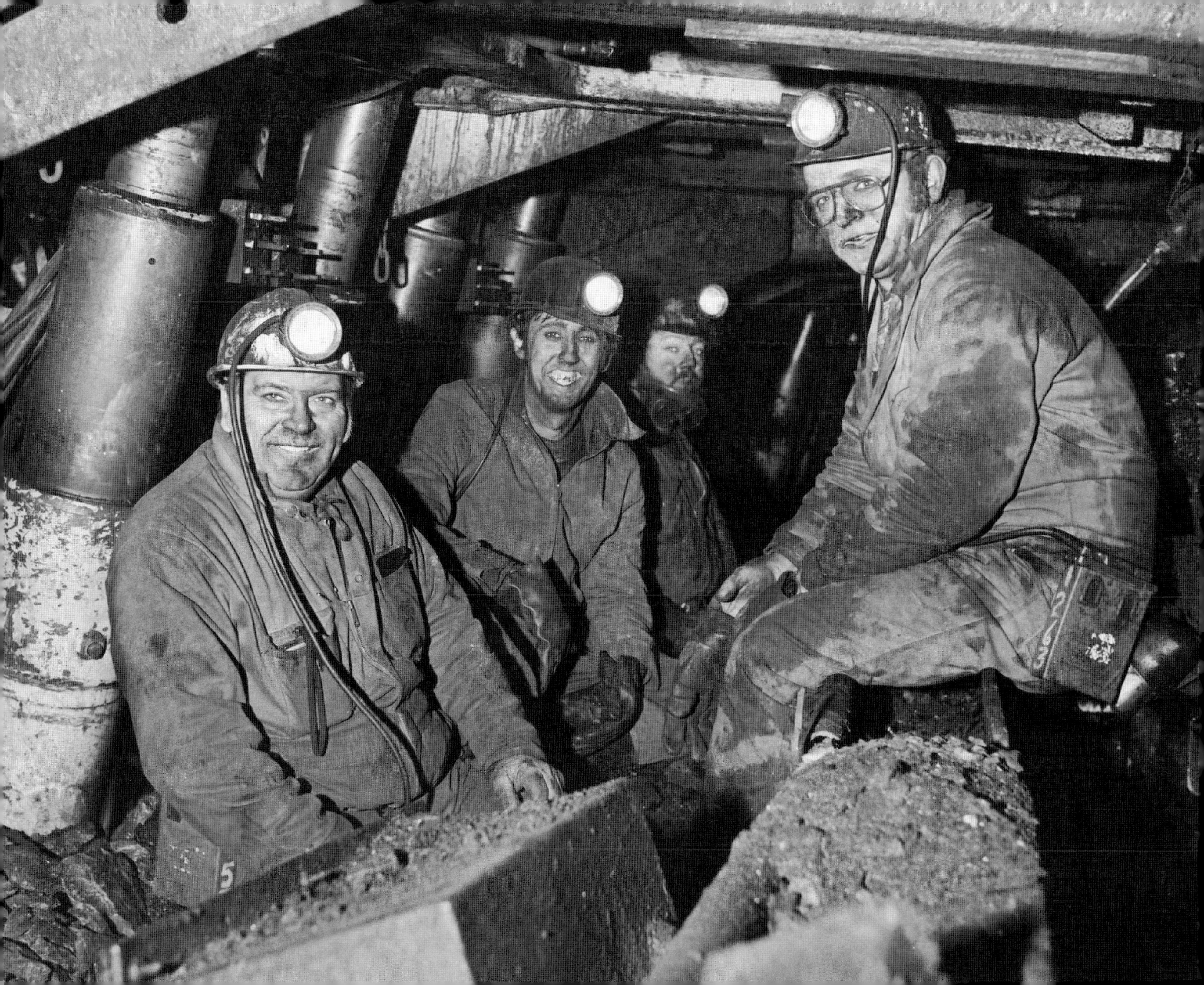

361-D-6 Prince Mine 1987

Inset Left 320-B-3 Prince Mine 1986

Inset Right 361-E-2 Prince Mine 1987

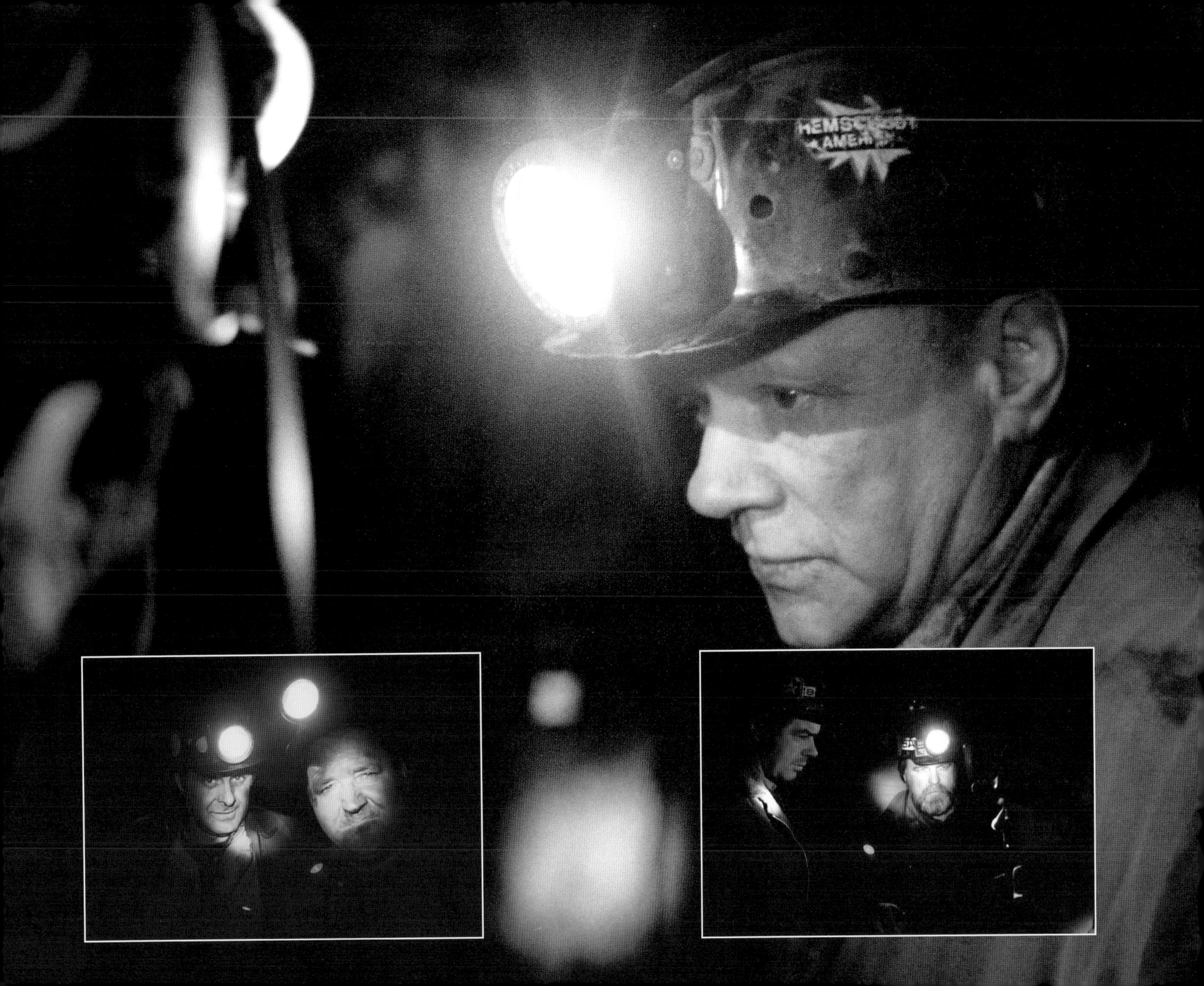

The checker.

We used to put a brass number in on the board (same number as our lamp) so they'd know who was down in the mine if something happened. You returned it at the end of your shift. If your number was on the board, you were in the pit.

185-A-24 Prince Mine 1983

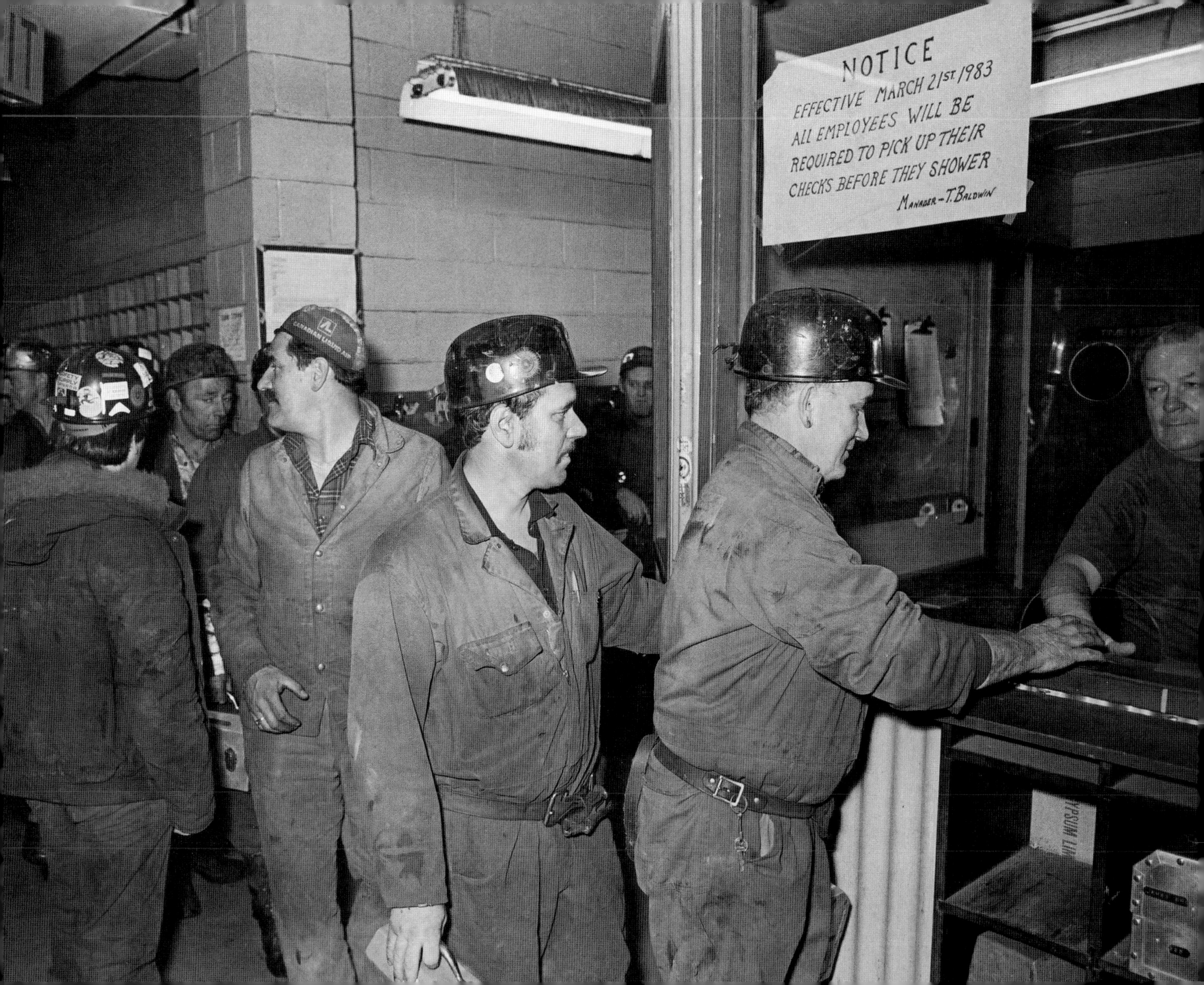
NOTICE
EFFECTIVE MARCH 21st 1983
ALL EMPLOYEES WILL BE
REQUIRED TO PICK UP THEIR
CHECKS BEFORE THEY SHOWER
MANAGER – T. BALDWIN

328-A-12 Donkin Mine 1986

DONKIN MINE
1984
1986

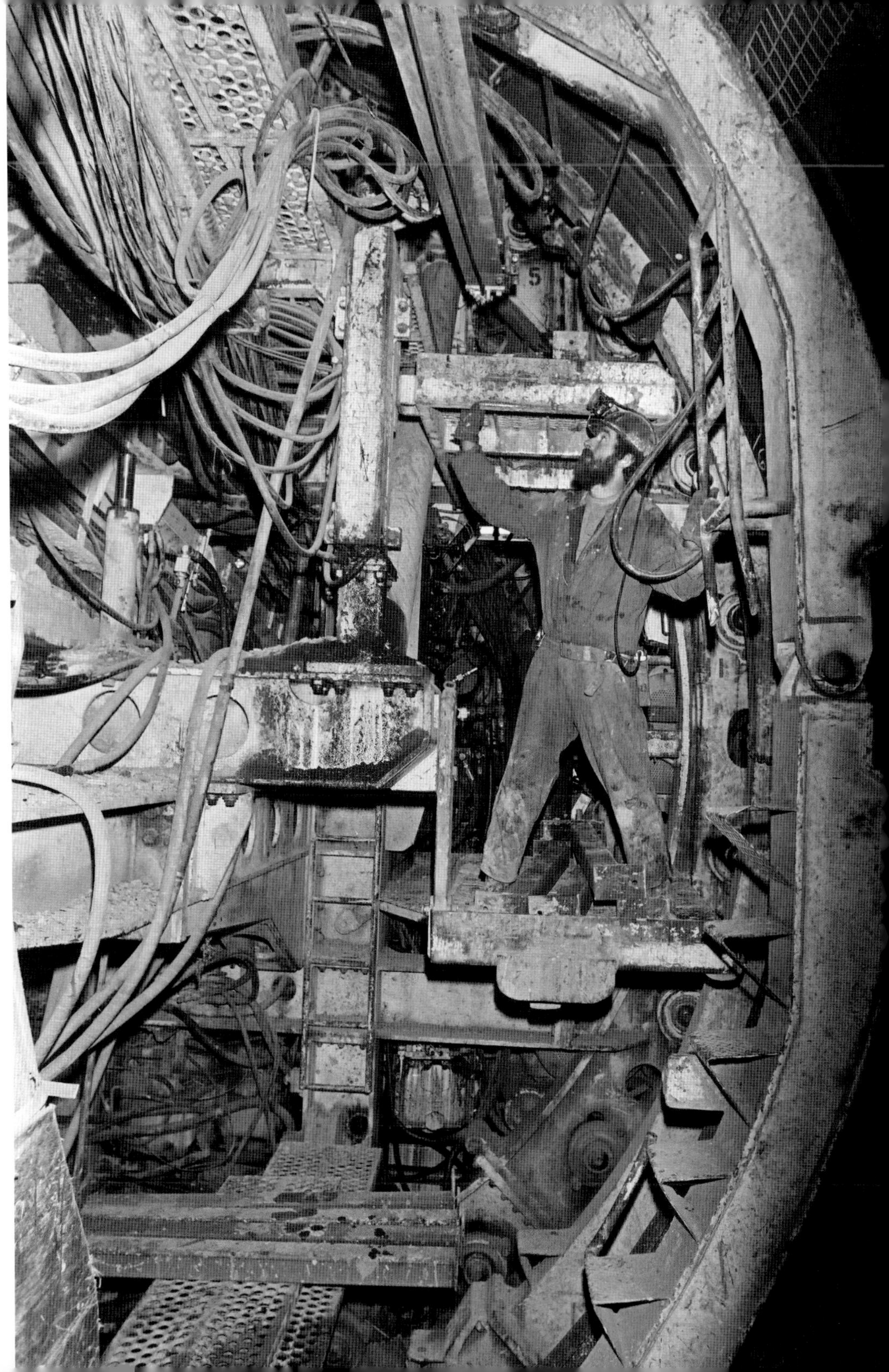

The tunnel boring machine—16 motors, 16 cutters. It cut right through the stone, just chewed it up 24 ft in diameter. It was 400 ft (approx. 120 m) when it was all together.

257-A-19 Donkin Mine 1984

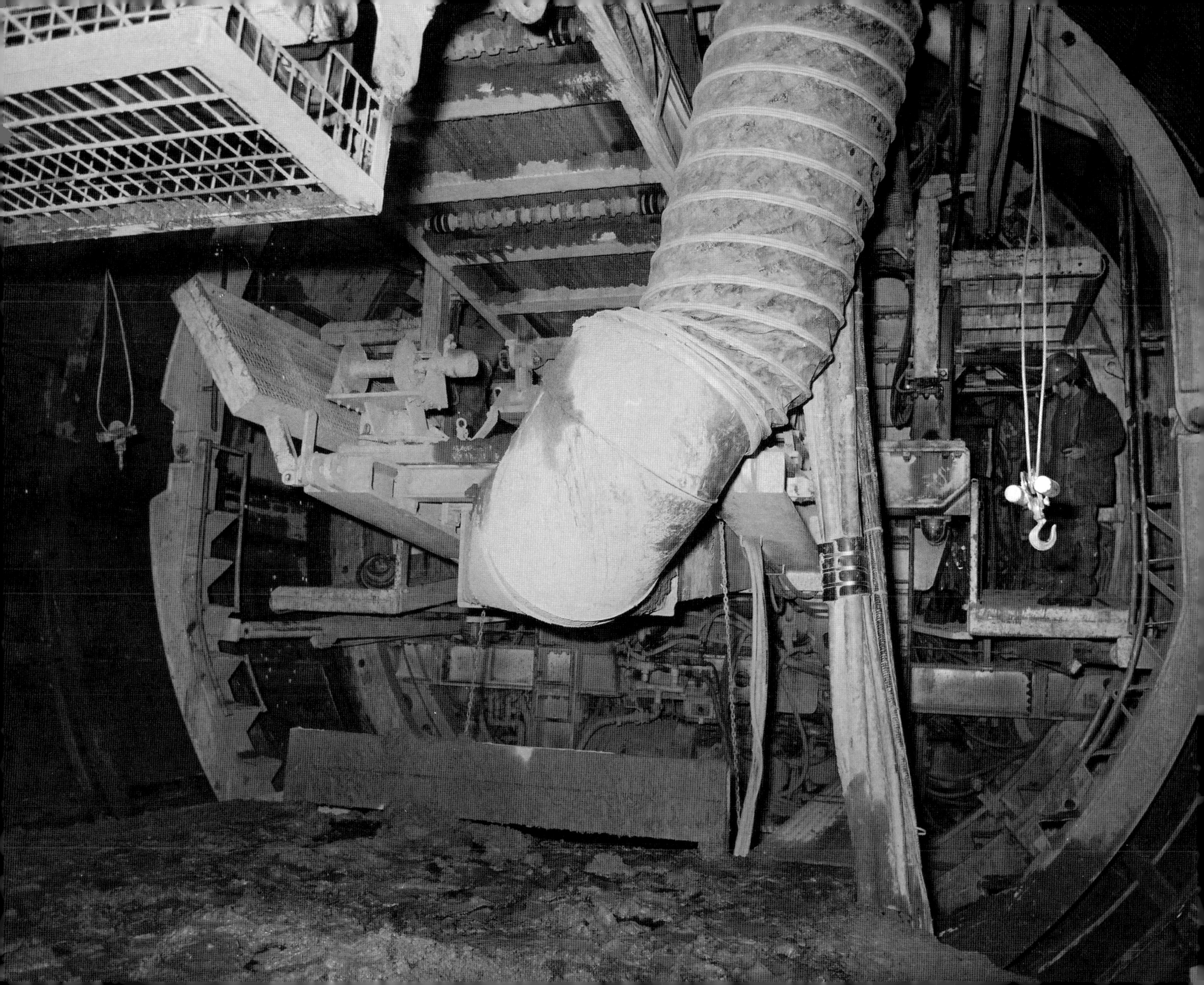

Operating the tunnel boring machine.

328-A-14 Donkin Mine 1986

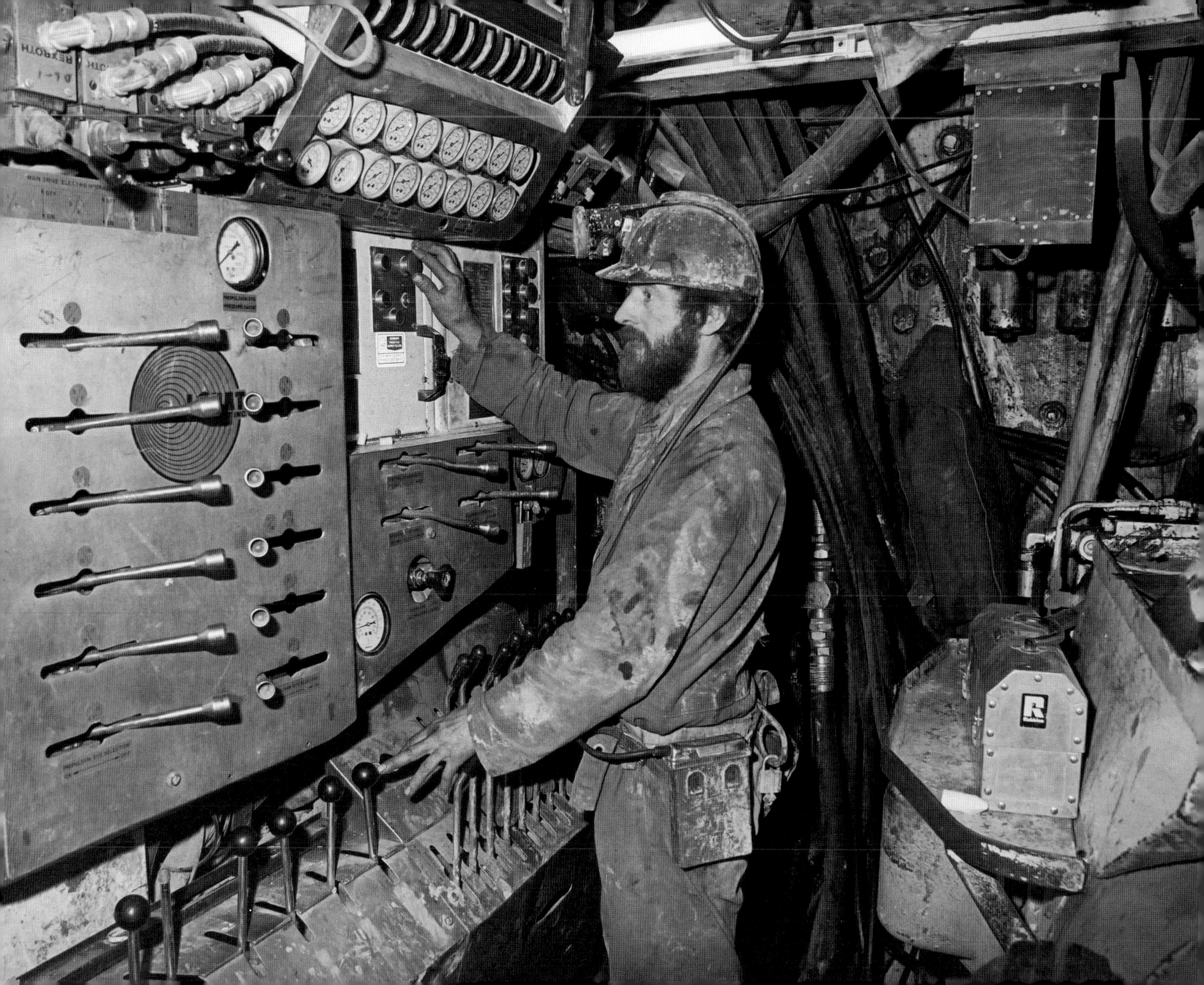
REXROTH

Donkin No. 2 slope. State of the art mine. Project to improve tunneling technology. The technology developed at Donkin is used worldwide, they used it to dig the Chunnel (English Channel).

257-B-3A Donkin Mine 1984

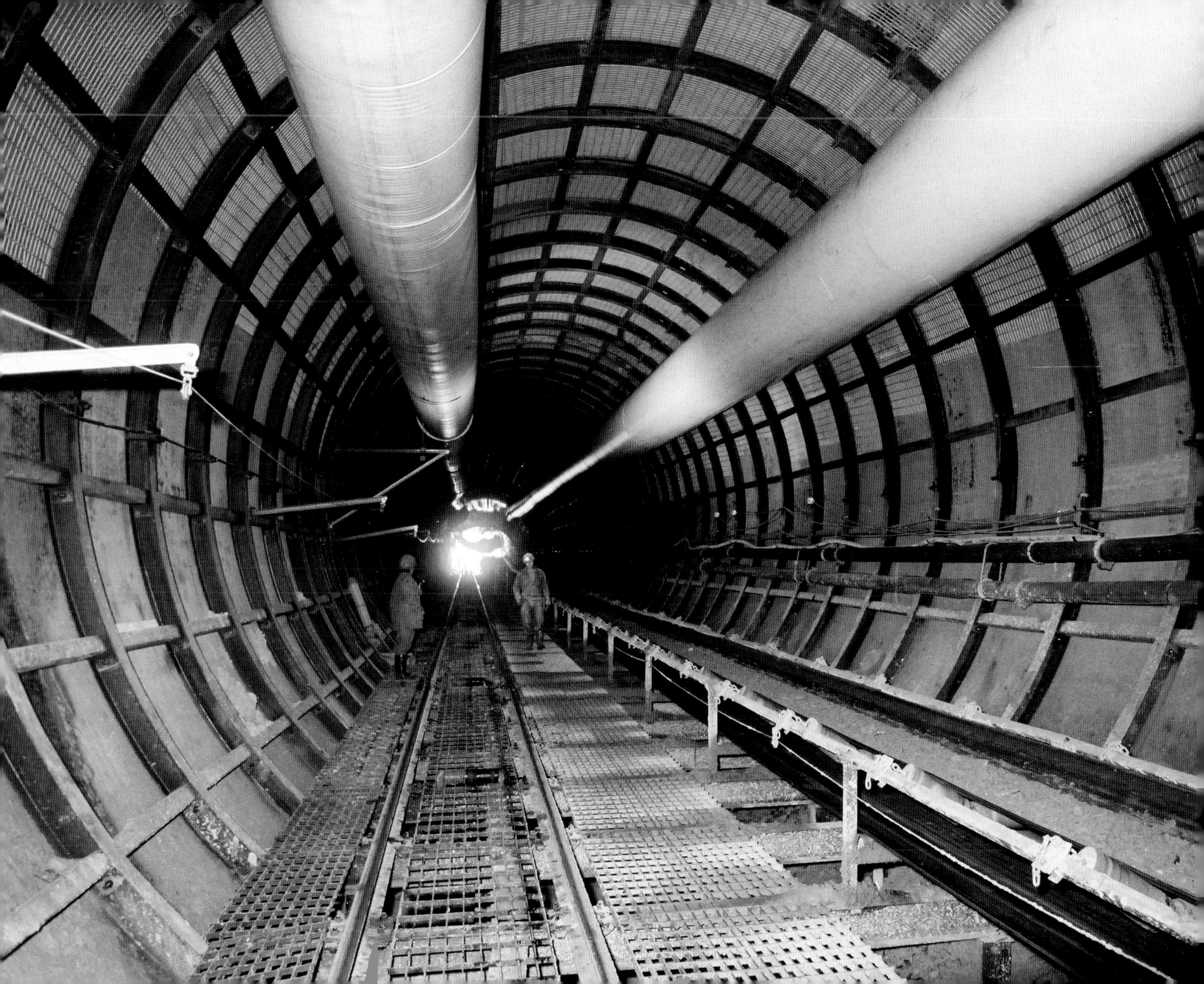

The monorail was used to hoist heavy equipment,
like this section switch.

57-A-11 Donkin Mine 1984

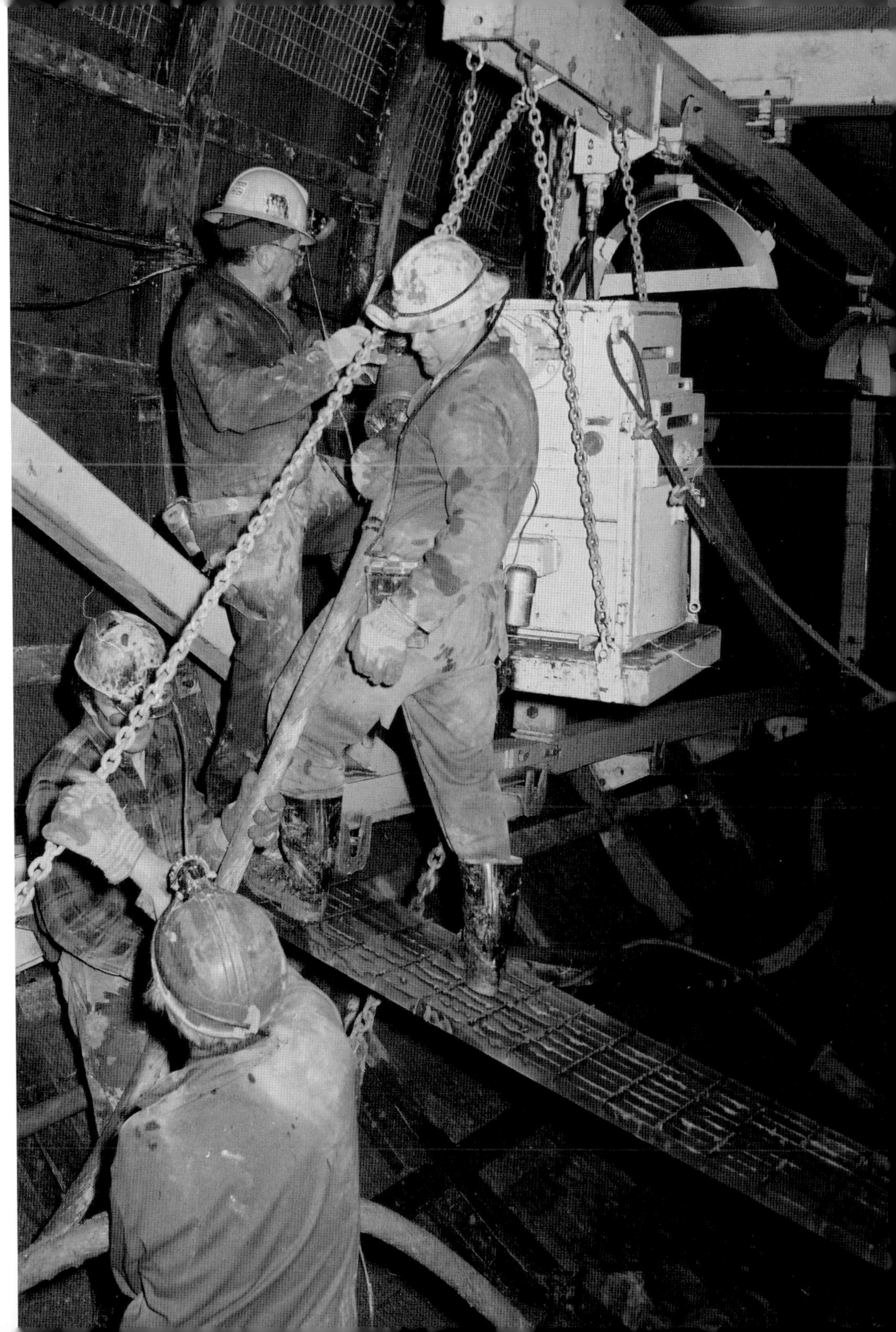

See the white hats? They didn't work for DEVCO. These men are operating air tuggers, working for the contractor. Note the fire extinguisher on the wall.

328-A-11 Donkin Mine 1986

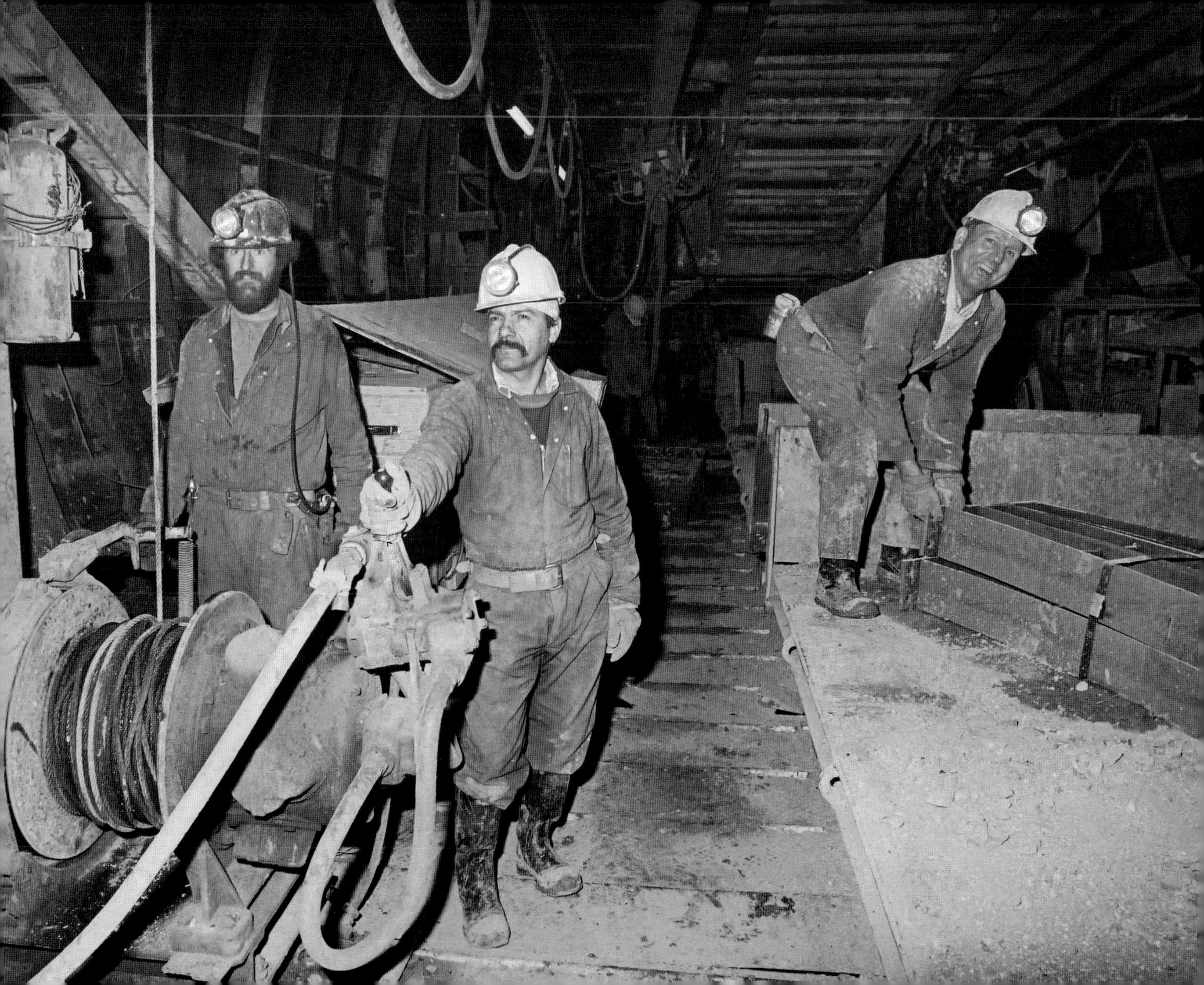

A switch relay. The motors are flame proof.

328-A-18 Donkin Mine 1986

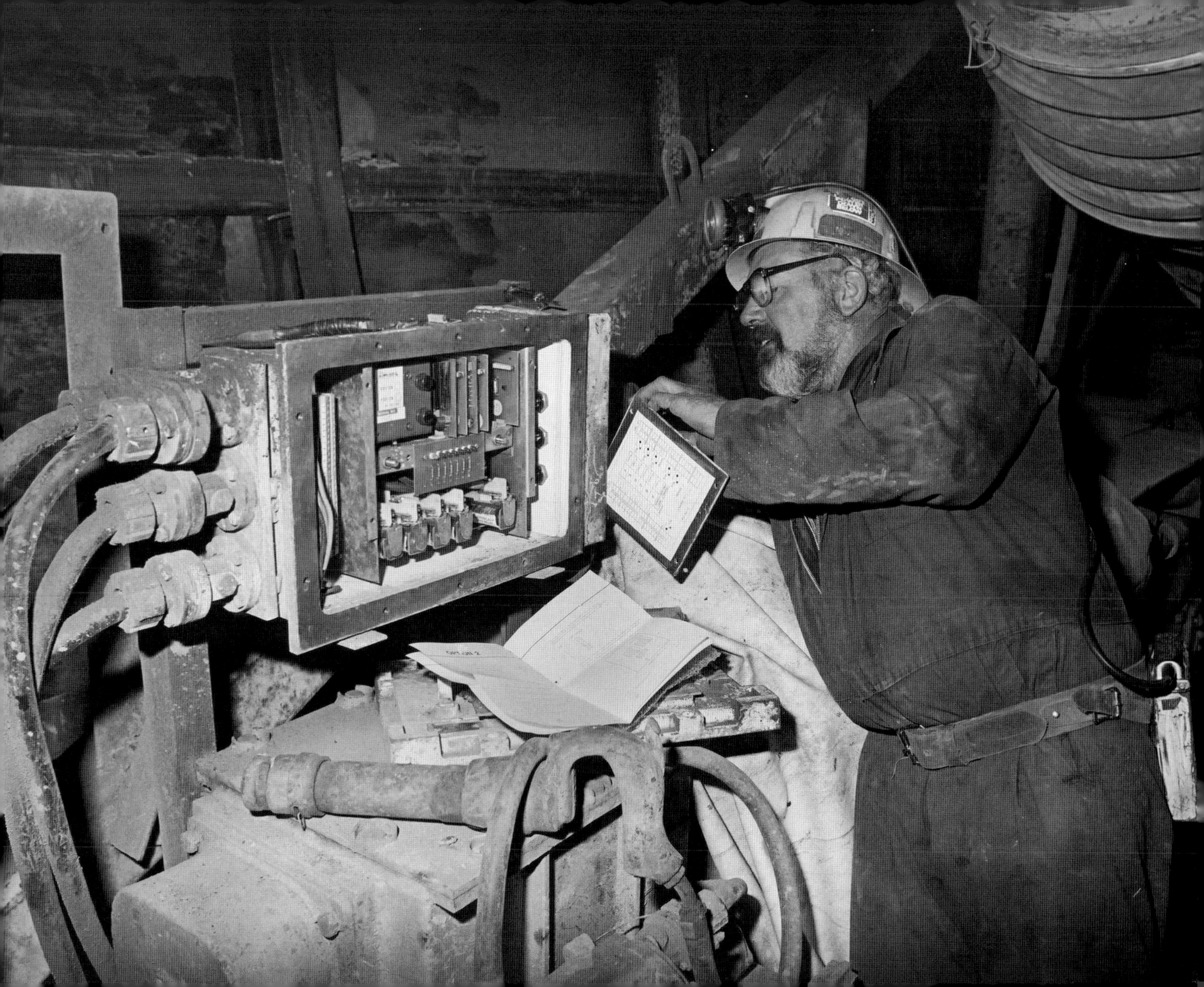

The Hunslet, a diesel fuel locomotive, used to haul material and men.

328-A-9 Donkin Mine 1986

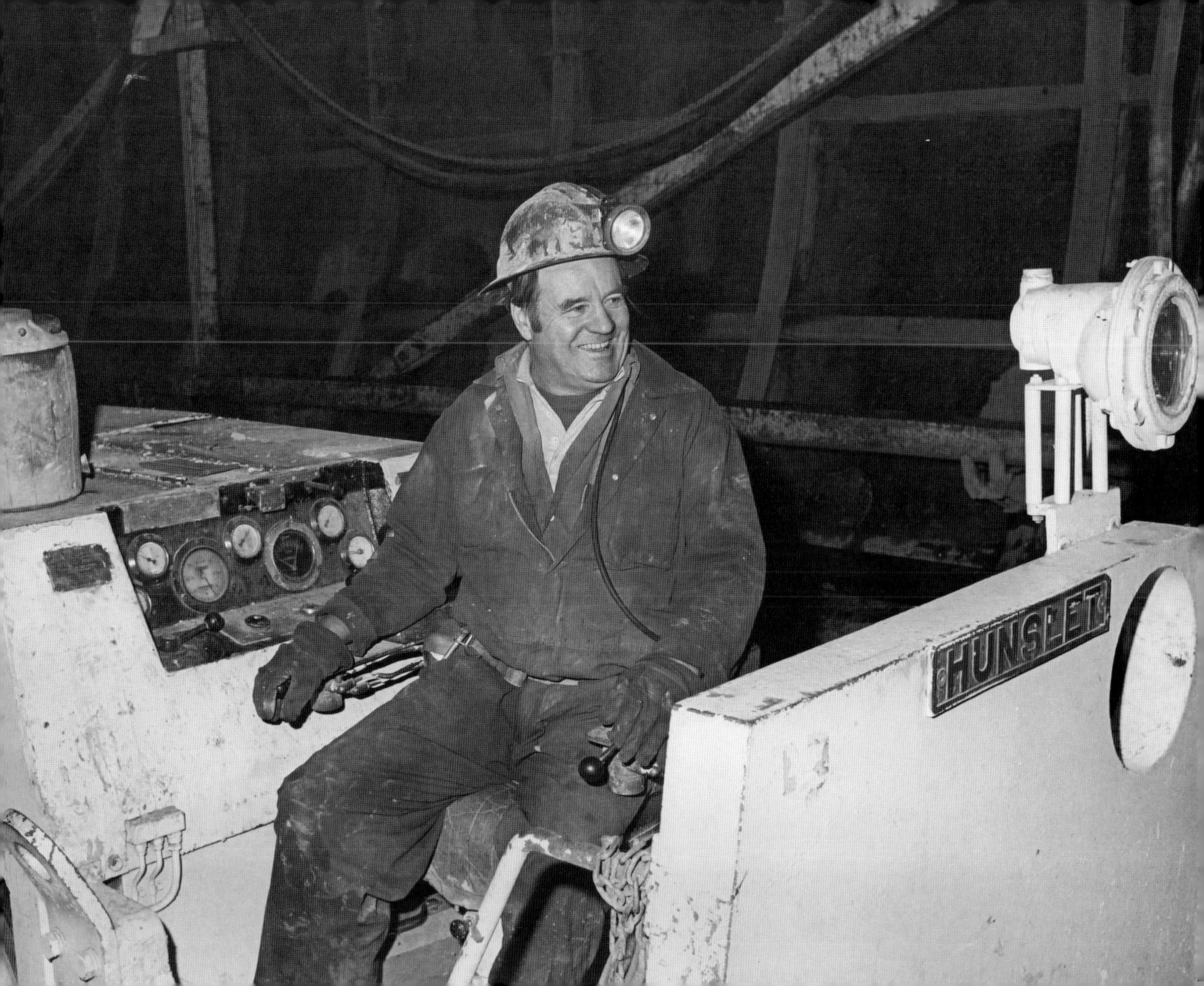
HUNSLET

Walking toward the surface on No. 3 slope, a 20% grade.

The rope on right was the communication system for the rake – 1 = stop; 2 = go down; 3 = go up. This was a drill and blast tunnel. Note the belt on left.

328-A-25 Donkin Mine 1986

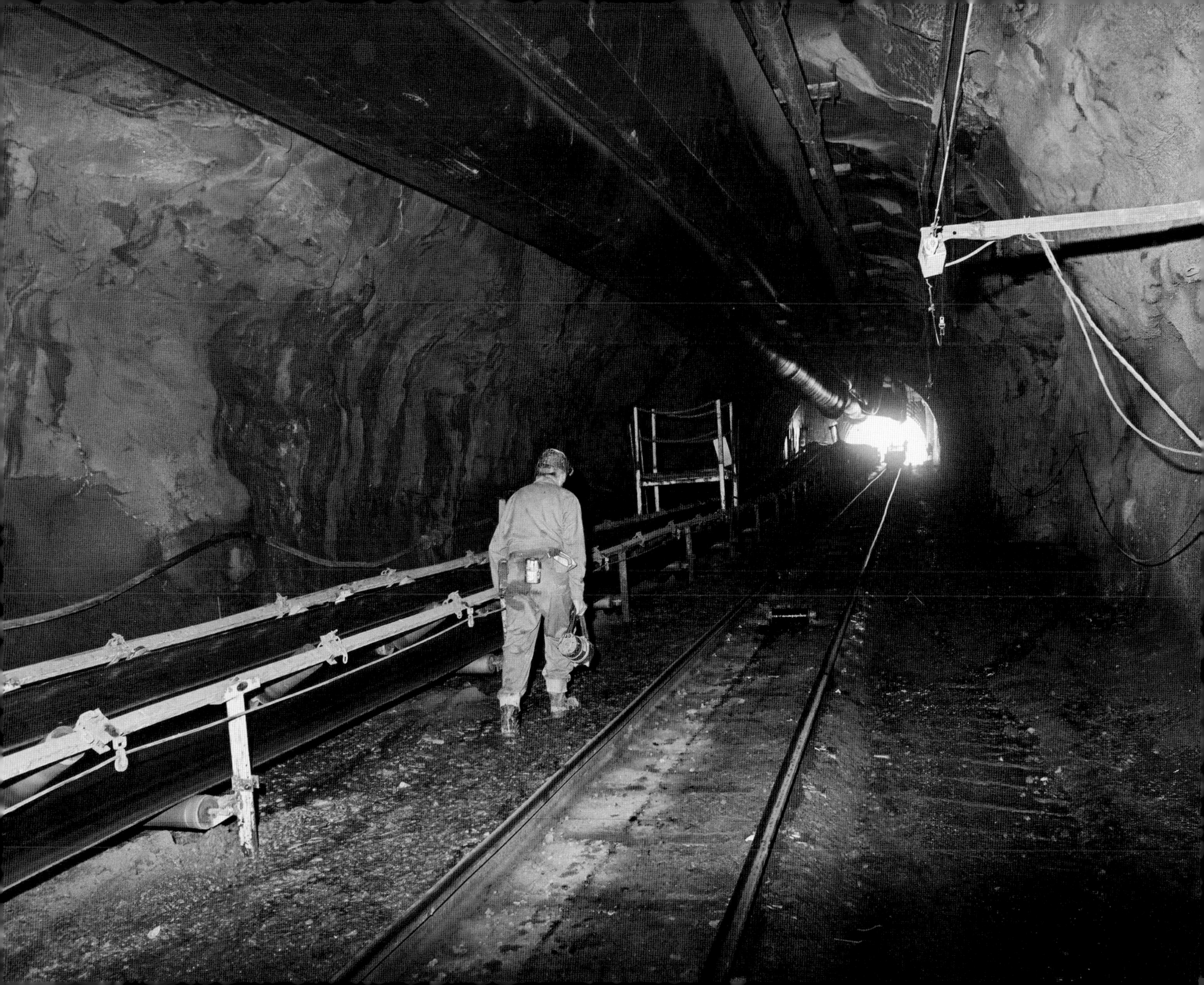

Coming out on No. 2 slope.

288-A-5 Donkin Mine 1984

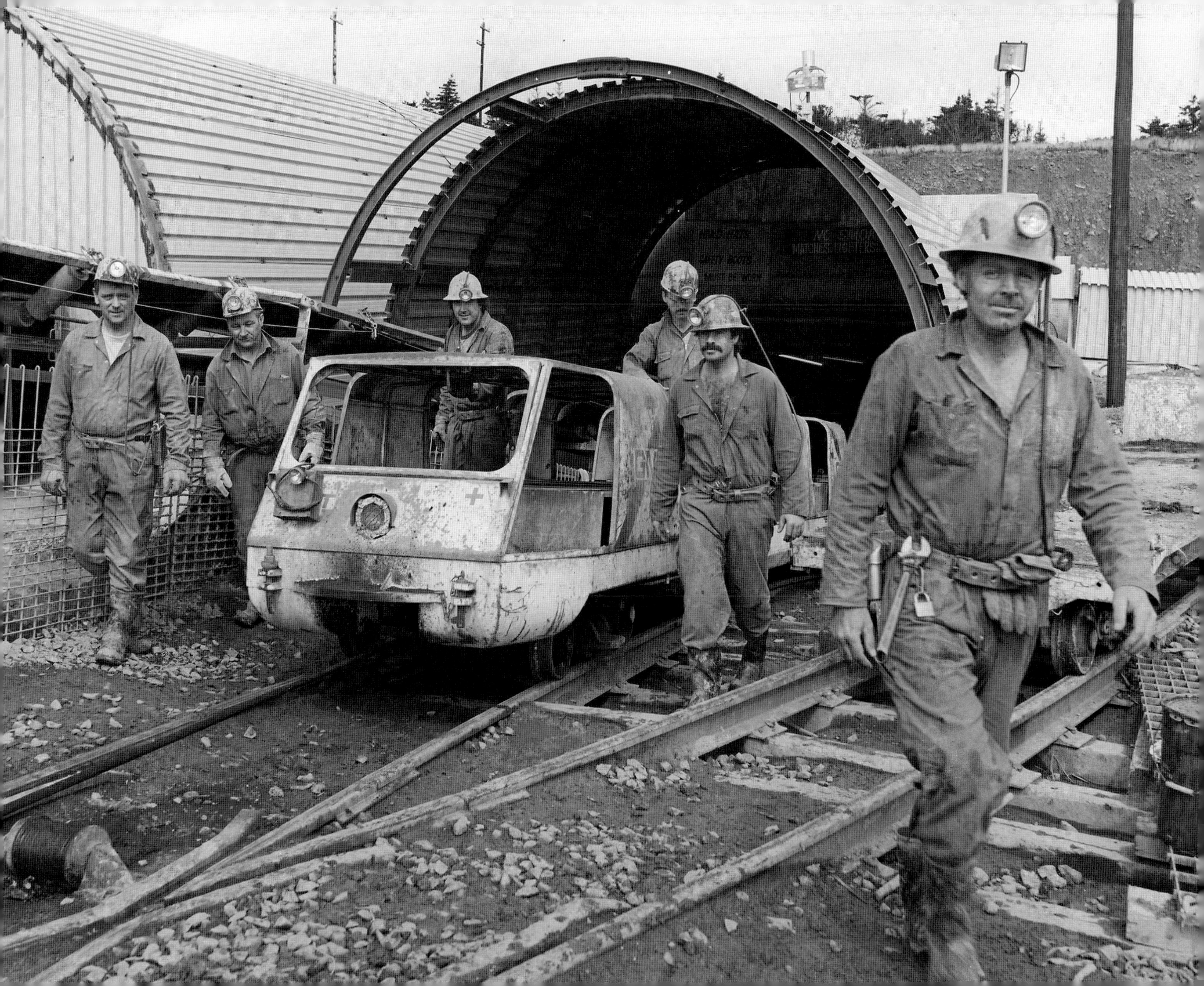

EPILOGUE

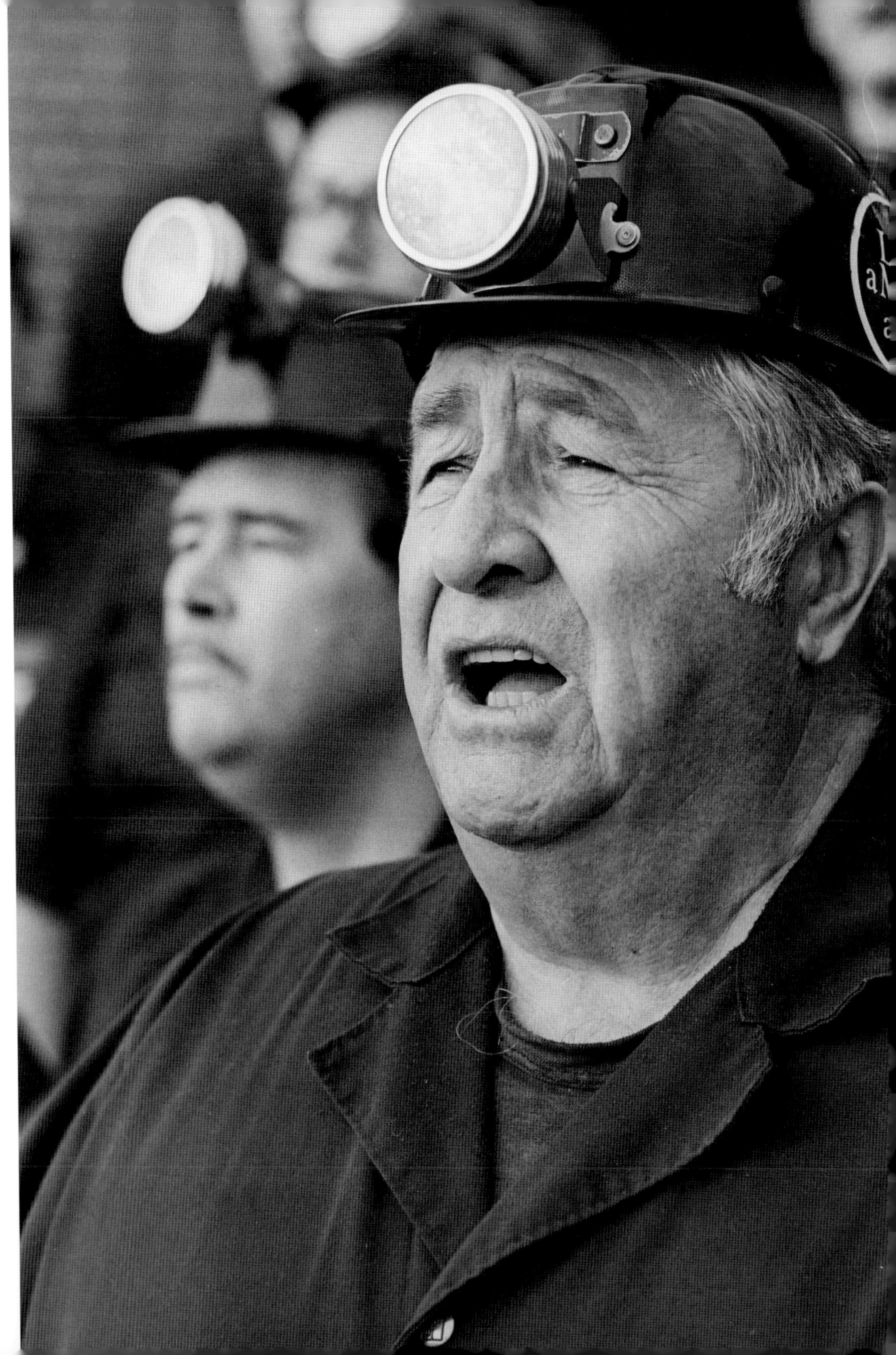

“I think everybody who worked in the mine did their work as fast as they could and as safe as they could, just to get back up on the surface.

Tape-2134 23 July 1984.

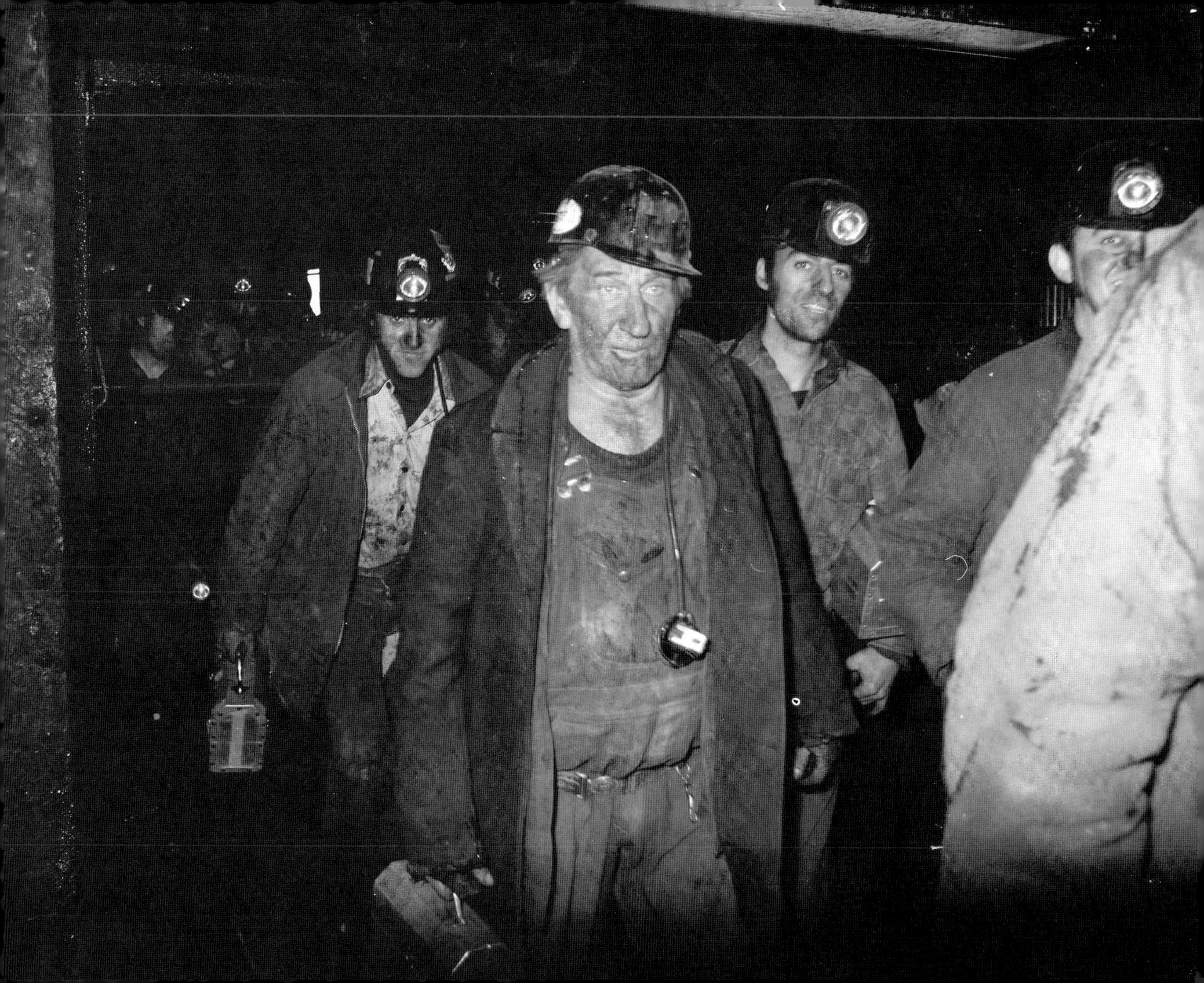

Because I was a regular visitor in the mines, I guess the miners and I established a relationship and respect for each other. As a result, I was regularly called upon to photograph the world famous miners chorale, the “Men of the Deeps.” I felt like an honourary member of the Men of the Deeps, as they got me to do photos of them every few years for different music CDs they produced. They are definitely a class act and great to photograph.

Owen

331-C-25 Men of the Deeps 1986

AFTERWORD

Almost exactly 100 years after it was opened, the oldest working coal mine in North America surrendered its last ton of coal on January 8, 1976, with little fanfare or public notice.

For 100 years, Princess played a crucial role in the economy of the Northside and the lives and fortunes of those who lived there. She was the source of income for as many of 1,000 men. Princess was referred to, as the "Winning Pit" or No. One Colliery.

Cape Breton's mining history goes back three centuries, when French explorers discovered coal near Sydney Harbour in 1672. The first commercial mine opened in 1720 and supplied the French encampment at the Fortress of Louisbourg.

By 1870, 21 collieries were churning out tons of coal to feed the eastern seaboard. "Some of the Fathers of Confederation thought Nova Scotia would be the industrial engine of the Dominion," said David Frank, a professor of Canadian history at the University of New Brunswick. The boom quickly ended after the Second World War.

When oil and gas took over it became difficult to sell coal. This was the beginning of the end for the Cape Breton coal mines.

The federal government held a royal commission and the commission said "future planning should be based on the assumption that the Sydney mines will not operate beyond 1981."

The Cape Breton Development Corp., or Devco, was created in 1967, assigned the job of phasing out the mines and diversifying an economy that had become reliant on one industry.

Then, just as Devco was gearing up to dismantle the industry, the 1973 oil crisis hit, sending oil prices skyward and producers looking for alternate energy sources. Cape Breton coal was king again.

New life was breathed into the coal mines and lured young men away from school and into the well-paid unionized work underground. DEVCO changed gears and opened several new mines to supply the world market.

Shortly before Princess colliery closed, I was lucky enough to be taken on a tour of the mine. What I remember most was the trip down on the man-rake. I'm not exaggerating when I say that this man-rake, which was just a small flat bed rail car with no top, flew down the mine slope and you had to lay back or risk loosing your head as the low ceiling zoomed by. I was thrilled to get some great photographs in Princess. One of my favourite photos was a close up of Michael Harrietta at the mine face, lit only by a fellow miner's helmet lamp.

Princess was the first submarine mine in North America. It was the first colliery in North America to mine coal beneath the bed of the Atlantic Ocean. Princess gave up 30 million tons of coal in her lifetime and claimed the lives of 150 men. She also formed the catalyst for a once-thriving steel industry on the Northside, at the time one of Canada's most modern steel plants.

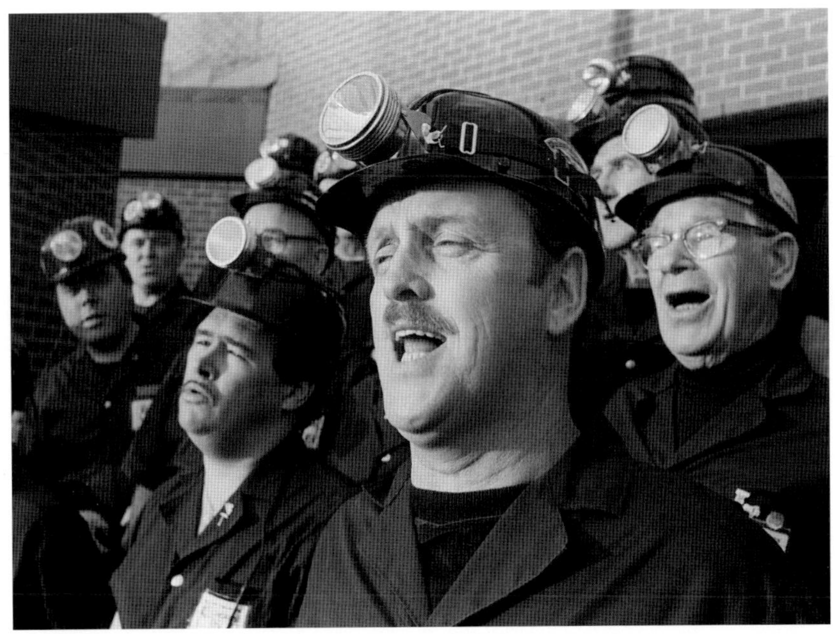

The Princess shaft dropped over 700 feet to reach the coal on the Harbour seam, which was the best of 27 coal seams around Cape Breton. This seam varied in thickness, from 4.5 feet to 9 feet (1.5 - 3 m).

Coal sold for $1.00 a ton in 1896, and miners received $1.00 per day. In 1976 coal sold for $50 per ton and miners earned $40 per day.

In 1883 there were 12 mines operating in Cape Breton and in 1892 Cape Breton mines produced one million tons of coal. In the early part of the twentieth century, coal mining was the only substantial industry in Nova Scotia. Cape Breton was producing 40% of the coal output of Canada in 1915.

The Sydney Mines Steel Mill operated for 20 years in competition with the nearby Sydney Steel Plant. To make for a more profitable operation, the two companies were merged in 1921 as part of the British Empire Steel Corporation (BESCO).

On December 6, 1938, disaster struck at Princess when the cable pulling the man-rake tangled and broke, setting loose the trip which derailed and piled up in a mass of twisted steel. Twenty-two men lost their lives and many more were injured. Following this disaster, a breaking device was installed for the first time on all man-rakes.

The outbreak of the First World War brought new demands for Cape Breton coal. Production increased and the collieries worked every day possible to help fill the need for coal. Many Cape Bretoners were even brought home from the war to work in the mines, so important was coal to the war effort.

Princess reached its peak production in the early 1960s. In 1963 she produced nearly 700,000 tons of coal and employed 1,100 men. In her lifetime, Princess gave up 30 million tons of coal.

The energy crisis of 1973 came too late for Princess colliery. Government decided that a new mine was the answer, and Prince mine in Point Aconi, opened during the last months of Princess's operation and that was 1976.

Then oil prices dropped and Cape Breton was stuck with new mines that had coal seams located several kilometers out under the ocean.

In 1999, Ottawa announced it would get out of the mining business for good. Close to 1,700 people lost their jobs. For the coal miners of Cape Breton, a centuries-old way of life had come to an end.

The closing of the Cape Breton coal mines and the Sydney Steel Mill affected the whole community. Personally, I decided it was time to make a career change. I began focusing more on my computer work in multimedia production and teaching. I also decided to go back to university. Since 1999, I completed by Business Degree and recently I received my Masters Degree in Business Administration (MBA) from Cape Breton University.

Having retired from Photography, and realizing I had photographed an important chapter in Cape Breton's history, I donated my collection of photographs to the Beaton Institute (Archives) at Cape Breton University (CBU). Through the publishing of this book, I am proud to be part of the 50th anniversary celebrations of the Beaton Institute. I also hope this book will serve to honour the proud Cape Breton miner, of which my grandfather, Frank Fitzgerald, was one.

I am also proud to have had a part in establishing the "Hope Scholarship Fund" at CBU. This fund is meant to assist mature students to go to school, in an effort to open the door to new opportunities. Because I believe that knowledge is power and education is the key to building a strong community, I am donating my royalties from the sale of this book to the "Hope Scholarship Fund."

Owen Fitzgerald, 2007

THE BEATON INSTITUTE

A gem of Cape Breton University and a true pillar in its commitment to culture and community turns fifty this year, the Beaton Institute, and its predecessor Cape Bretoniana, serves as the regional archive for Cape Breton Island and as the institutional archive for Cape Breton University.

In 1957 Sister St. Margaret of Scotland (Sister Margaret Beaton), in her role as Librarian of Xavier Junior College, recognized that many documents of historic and cultural significance to Cape Breton Island were being lost due to neglect and the lack of an appropriate repository. In a pro-active response, Sr. Beaton began collecting Cape Breton related documents and records with the objective of preserving materials of historic significance for future scholars.

Today, the Beaton Institute is recognized as having a rich regional collection including: 3,000 manuscript collections; 60,000 images; 2,500 sound recordings; 1,500 video cassettes and film reels; 1,500 reference books; 2,000 maps and plans. This collection is made available in the modern facility of Cape Breton University to students, scholars and the local and global research community.

The Beaton Institute collects and preserves the social, economic, political and cultural history of the Island and through this mandate supports and fosters research related to Cape Breton—its people, its industry, its history and its rich cultural heritage.

This book was initiated as one of a series of projects to commemorate the 50th Anniversary of the Beaton Institute. This publication has used various sources from the archive's holdings including the photographic, audio-visual and manuscript collections.

An archive acquires and makes available to the public documents of enduring or continuing value so that future generations will have access to the records of their ancestors, community and society. Photographs are a rich source of historical information—and this publication now offers wider access to the photographic record of mining in Cape Breton Island.

Catherine Arseneau
Manager, Beaton Institute

ACKNOWLEDGEMENTS

Many thanks to the many people who shared their time and insights.

Miners Focus Group:
Stephen Deveaux, who worked from 1980-2003 at No. 26 Colliery with the Engineering Department, servicing all mines and support sites, and who was responsible for underground transportation.

Stephen Woods worked at the Princess Colliery from 1977-2001. While growing up in Sydney Mines, he and his friends played at the Princess Colliery site. He was part of the effort to extinguish fires at No. 26 Colliery in 1984.

Robert MacDougall worked from 1977-1984 at No. 26 Colliery; 1984 and 1999-2001 at Prince Mine; 1985-1999 at Phalen Mine; 2004-2006 at Grande Cache Coal, Grande Cache Alberta; and 2006-2007 Donkin Mine.

Donnie Campbell worked at No. 26 Colliery, Lingan, Princess Mine, Phalen and Donkin.

Abbie Michalik worked at No. 26, No. 4, No. 20 and Phalen Mines.

Neil MacKinnon worked at Lingan, Phalen, Prince and Donkin Mines.

Beaton Institute Student Archival Research Assistants: Alicia Lamey and Kelly Krawchuk

Owen Fitzgerald for generously donating his images to the Beaton Institute and for collaborating on this project.

Staff of the Beaton Institute and all donors to the Beaton Institute's collection—without their generous donations our knowledge of our past would be greatly diminished.

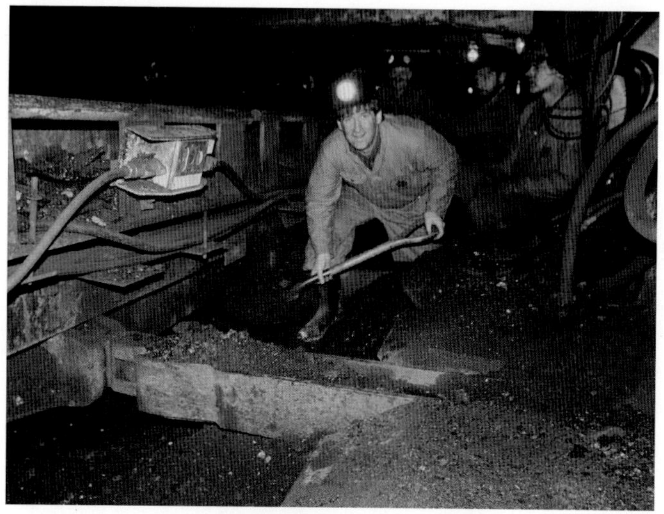

379-2-10 Phalen Mine 1987